Stomata

Colin M. Willmer
Department of Biology, University of Stirling

WITHDRAWN

Longman *London and New York*

MILTON CAMPUS LRC

Longman Group Limited
Longman House
Burnt Mill, Harlow, Essex CM20 2JE,
England and Associated Companies throughout the World.

Published in the United States of America
by Longman Inc., New York

© Longman Group Limited 1983

All rights reserved. No part of this publication may be reproduced, stored in a retrieval system, or transmitted in any form or by any means, electronic, mechanical, photocopying, recording, or otherwise, without the prior permission of the Copyright owner.

First published 1983

British Library Cataloguing in Publication Data

Willmer, Colin M.
Stomata
1. Stomata
I. Title
581.1′21 QK873

ISBN 0-582-44632-5

Library of Congress Cataloging in Publication Data

Willmer, Colin M. (Colin Michael), 1942-
Stomata.

Bibliography: p.
Includes index.
1. Stomata. I. Title.
QK873.W54 581.1′2
ISBN 0-582-44632-5

Printed in Singapore by
Huntsmen Offset Printing Pte Ltd.

84-4400/MCL

Contents

Preface

Over the last decade enormous advances in our understanding of the functioning of stomata have been made. Today, the overall pattern of events in stomatal functioning is generally understood although some of the finer details of the mechanism remain to be elucidated. However, the mechanism whereby changing CO_2 levels within a leaf control stomatal movements remains one of the biggest puzzles.

The excellent text *Physiology of Stomata* by Professors H. Meidner and T. A. Mansfield, is over twelve years old and inevitably sections of it are now outdated. Although numerous fine reviews on various aspects of stomata have been written in the meantime, none could be considered as complete, compact appraisals of the subject and thus to have superseded the Meidner and Mansfield book.

Now is therefore an appropriate time to review the status of our knowledge about stomata.

This book encompasses all aspects of the subject except the physical processes involved in the exchange of gases between a leaf and its environment. This topic is dealt with very adequately in a variety of other books and reviews and the subject matter, with minor differences of opinion, remains much the same today as it was perhaps twenty years ago. A description of how to measure stomatal apertures and the use of porometers has been restricted to a final chapter which also deals with experiments which can be carried out by students. Again, there are numerous texts and papers which deal in greater detail with the theory and use of porometers than that described here.

In a text of this type, which is not intended to be a research treatise but a more general, though detailed appraisal, of the subject intended for undergraduate and postgraduate students specialising in plant sciences, references must be limited.

Choosing the limited number of references was difficult. I have tended to quote the more recent references, important and major articles of the past, and review articles. Inevitably, however, some statements remain unsupported by a reference. Also, in attempts to keep the book as current as possible, I have drawn on information from a number of unpublished works.

Where contentious issues arise I have attempted to state each view but space has not always allowed me to discuss the detailed evidence supporting each view. The reader, in some cases, is left to read the original reports to decide the most convincing line of argument.

I have also tried to be specific about the plant species used in a particular experiment because, although there are certain basic features and processes concerning stomatal functioning and behaviour which are common to all species, there may well be variations on this common theme depending on the species.

Colin M. Willmer
Stirling (1982)

Author's acknowledgements

I am particularly indebted to Professor D. A. Baker, who edited my manuscript and gave valuable advice. A special thanks goes to Professors J. E. Varner and H. Meidner, both of whom read much of my manuscript and a number of improvements resulted from their criticisms.

Thanks are also due to the following: Dr W. H. Outlaw Jr of the Department of Biological Sciences, Florida State University, Tallahassee; Drs P. Brown, J. Maniotis, Barbara Pickard and others within the Department of Biology, Washington University, St Louis, all of whom helped me in their individual ways; colleagues within the Department of Biology, University of Stirling, for their comments and opinions about various points; T. Forrest, W. Struthers and L. Taylor of the Department of Biology, University of Stirling for EM and photographic work; numerous people, whose names are mentioned in the figure or table legends, who gave me permission to use their published and unpublished data. I also thank the following bodies who gave me permission to reproduce some of their published material: the American Society of Plant Physiology, Springer-Verlag, Cambridge University Press, Blackwell Scientific Publications Ltd, the Institute of Biology, London, and the National Research Council of Canada.

However, I am completely responsible for the views expressed in this book, for the presentation of its contents and any errors that may be found therein.

I would also like to acknowledge that this book would not have been written without the influence of Professor T. A. Mansfield, who, in the very beginning, gave me the opportunity to study and research into stomata and who initiated my interest in the subject.

Publisher's acknowledgements

We are indebted to the following for permission to reproduce copyright material:

The American Society of Plant Physiologists and the authors for our Figs. 5.3 & 5.4 from Figs. 4 & 5 (Hsiao *et al* 1973) and 5.5 from Fig. 1 (Cockburn *et al* 1979); Blackwell Scientific Publications Ltd for our Fig. 5.7 (Watts & Neilson 1978); Cambridge University Press and the author for our Fig. 3.7 from Fig. 7 (Palevitz 1981); Institute of Biology and the author for our Fig. 5.8 from Fig. 10 (Mansfield 1971); LI-COR for our Fig. 8.2 from p 8 (*LI-COR Catalogue* 1981); Professor H. Meidner for our Fig. 8.4 from Fig. 2.4b (Meidner 1968); National Research Council of Canada and the authors for our Figs. 2.12 from Fig. 30 (after Dayanandan & Kaufman 1973), 3.13 from Fig. 2 (Allaway & Setterfield 1972), 3.15 from Fig. 8 (Pallas & Mollenhauer 1972), 6.4 from Fig. 3 (Allaway & Mansfield 1970) and Table 3.1 from Table 1 (Allaway & Setterfield 1972); New Phytologist for our Fig. 5.9 adapted from Fig. 1 (Martin & Meidner 1971); Springer-Verlag Heidelberg and the authors for our Figs. 3.6 from Fig. 36a–f (Palevitz & Hepler 1974a), 4.6 from Fig. 6a–e p 394 & 5.2 from p 388 (Raschke 1979), 5.6 from Fig. 2 (Tenhunen *et al* 1980) and Table 6.1 from p 160 (Tucker & Mansfield 1971); Washington State University for our Figs. 8.1 & 8.5 from Figs. 3.1c & 3.18 (Washington State University 1975).

1 Introduction

Stomata are pores formed by a pair of specialised cells, the guard cells, which are found in the surface of aerial parts of most higher plants and which can open and close to control gas exchange between a plant and its environment. Their purpose is to act as portals for entry of CO_2 into the leaf for photosynthesis and exit of water vapour which may be used for evaporative cooling of the leaf. The transpiration stream may also facilitate uptake and transport of salts necessary for the nutrition of the plant.

Stomata appeared in plants at least as early as the Lower Devonian period (about 390 milllion years ago), occurring in various genera including *Zosterophyllum*, *Drepanophycus*, *Rhynia* and *Asteroxylon*. They probably developed simultaneously with a vascular system and a cuticle, these three features being chiefly responsible for the emergence and development of large terrestrial plants (Chaloner 1970).

The word stoma (pl. stomata) is Greek for mouth, and Link and de Candolle (de Candolle 1827) jointly claim to be the first to have called the pores by that name.

Conventionally, a stoma is accepted to mean the pore and the surrounding pair of guard cells. A stomatal complex refers to the guard cells and neighbouring subsidiary cells.

Soon after the invention of the simple microscope the first published observations of stomata were made by Malpighi (1628–94) in his manuscript, *Anatomia Plantarum*, published in 1674. He could make nothing of their function, however. Malpighi was better known for his studies of the skin and circulation system in animals, although some of his major scientific contributions are on plant anatomy.

Soon afterwards Nehemiah Grew (1641–1712) made similar observations and drawings of stomata to those of Malpighi and his series of lectures given to the Royal Society in 1680 when Sir Isaac Newton was president of the society were published in 1682. Figures 1.1 and 1.2 are reproductions of some of his drawings. Grew called stomata 'orifices' or 'pass-ports' and believed that they opened up into xylem vessels which he called 'aer vessels'. He considered that

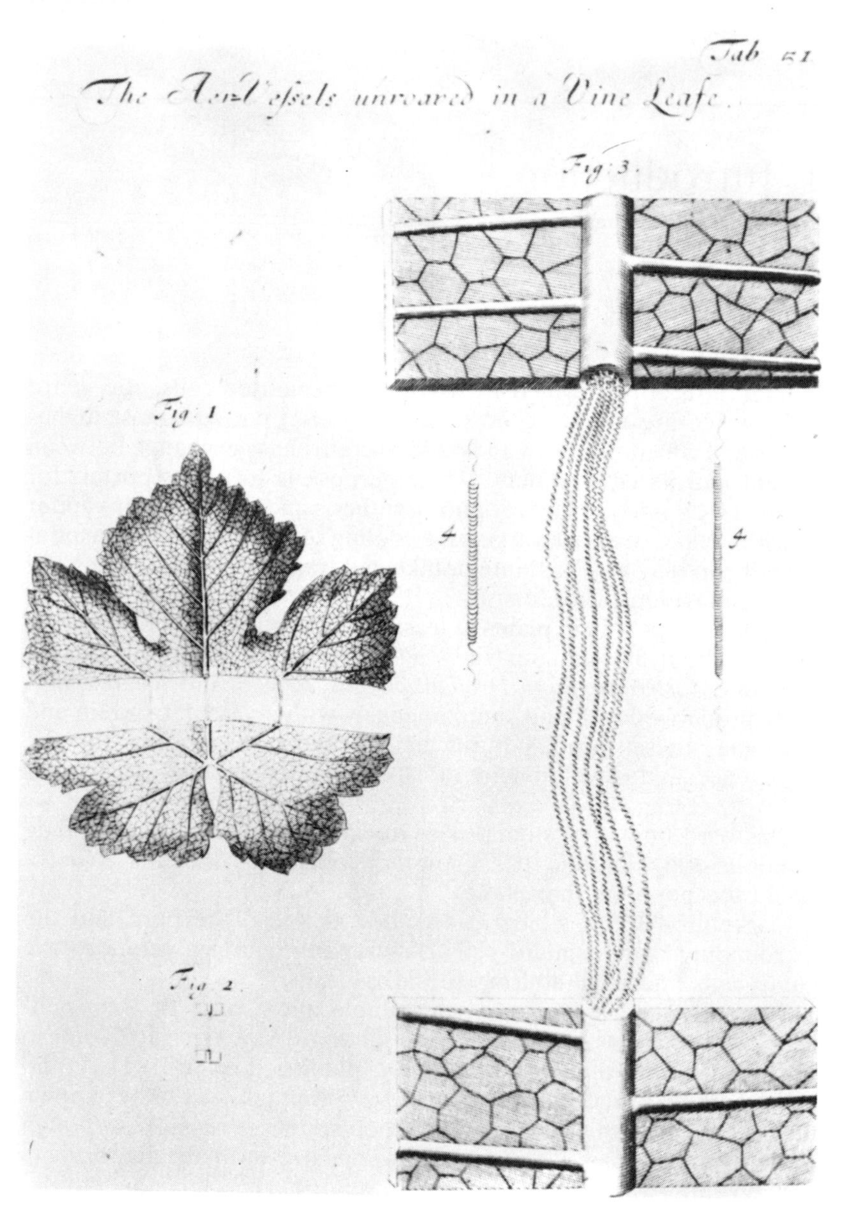

Figure 1.1 'Aer vessels' in a vine leaf (from Grew 1682)

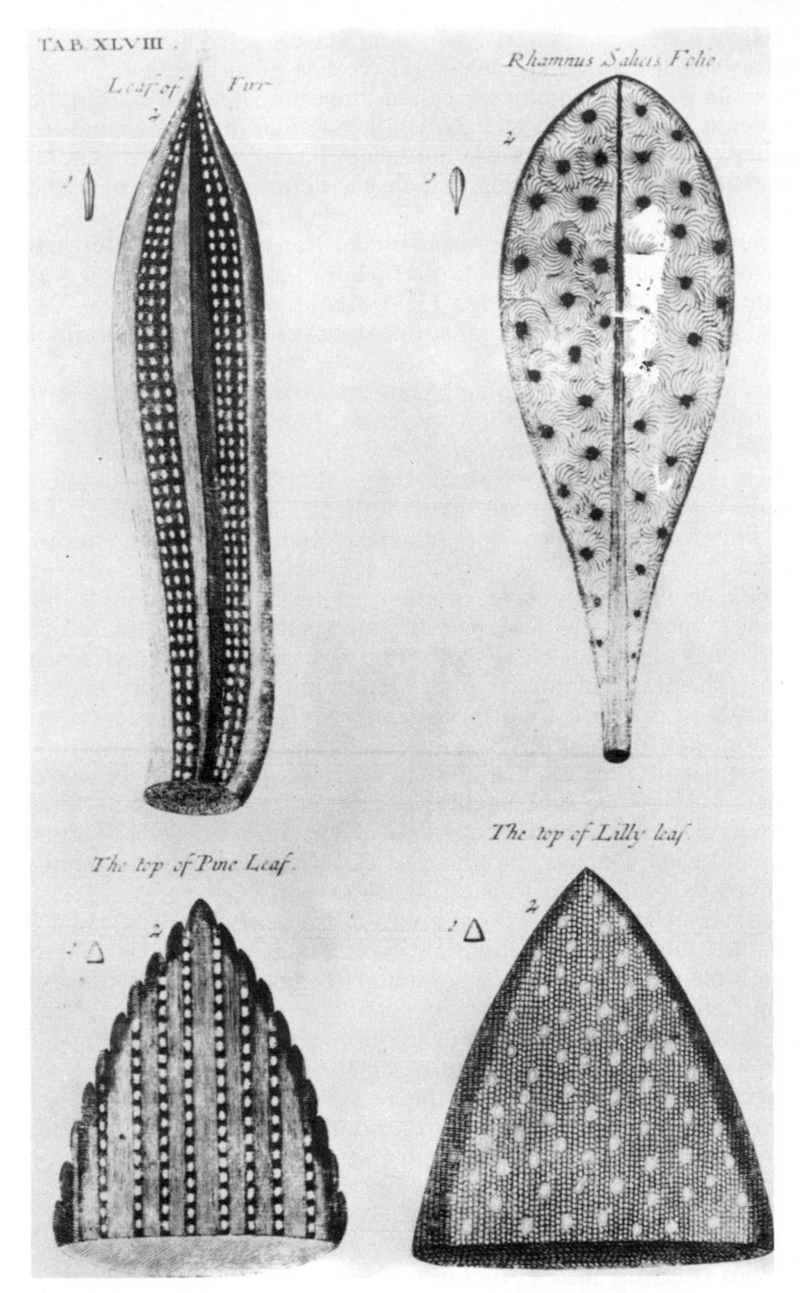

Figure 1.2 Low power drawings of leaf surfaces showing the patterns and locations of stomata (from Grew 1682)

the whole system of stomata and xylem vessels acted as an aeration pathway much like the spiracles and trachea of insects.

Stomata were still commonly called 'breathing holes' well into the nineteenth century and it was not until 1841 that the experiments of Williams, a lecturer in forensic medicine, finally dispelled a general view that a membrane, across which gas diffused, existed over the pore.

A number of eminent scientists studied transpirational water loss by plants including the Rev. Stephen Hales (1677–1761), but it was not until the work of Dutrochet (1776–1847), von Mohl (1805–72), Sachs (1832–97) and others that studies began into the mechanism of how stomata functioned.

Since then interest in stomata has increased enormously and a vast amount of information on their structure, function and ecology has been accumulated.

There are numerous reasons for the great interest in stomata. Foremost is that stomata are involved in the control of two of the most important plant processes, namely photosynthesis and transpiration.

Stomata control 95 per cent, or more, of the CO_2 and water vapour exchange between the leaf and the atmosphere. They therefore control rates of photosynthesis and transpiration by plants and, since photosynthesis is a major factor in determining rates of dry matter accumulation, stomata are of prime importance in any consideration of factors controlling crop yield.

Stomata and stomatal complexes are also proving to be ideal subjects for studying fundamental aspects of cell development and differentiation. For example, our knowledge about how cell polarity arises and what controls the plane of cell division has been greatly improved by studies on stomatal development.

From a biochemical viewpoint guard cells are also proving to be a fascinating subject for investigation. Not only are guard cells unique amongst plant cells because of the metabolic machinery they possess but they are also able to sense very small changes in CO_2 concentrations and respond by setting in motion a series of biochemical reactions which will result in the stomata opening or closing.

Guard cells are isolated from the rest of the plant body (but *see* Ch. 3) and, except for certain reproductive cells where genetic material needs to be separated, may be the only cells which are not connected to other cells via plasmodesmata.

Further reading and references

Chaloner, W. G. (1970) The rise of the first land plants, *Biol. Rev. Cambridge Phil. Soc.*, **45**, 353–77.

De Candolle, A. P. (1827) *Organographie Végétale*, Vol I. Paris.
De Saussure, M.-B. (1762) *Observations sur l'ecorce des feuilles et des petames*. Genève.
Grew, N. (1682) *Anatomy of Plants*. London.
Hales, S. (1727) *Vegetable Staticks*. London.
Malpighi, M. (1674) *Anatomia Plantarum*.
Rehfous, L. (1917) Etude sur les stomates, *Bull. Soc. Bot. Genève*, Ser. 2, **9**, 245–50.
Von Mohl, H. (1856) Welche Ursachen bewirken die Erweiterung und Verengung der Spattöffnungen?, *Bot. Ztg*, **14**, 697–704, 713–20.
Williams, T. (1841) On the structure and uses of the stomata, *Microscopic J. and Structural Record*, Vol. 1 118–21.

2 Morphology and anatomy of stomata

2.1 Location and occurrence of stomata

Stomata are found in all higher plants down to the evolutionary level of the phylum Bryophyta. Within the Bryophyta stomata are present in the capsules (sporophyte generation) of the Bryopsida (mosses) and Anthocerotopsida (horned liverworts). Stomata are not found in the Marchantiopsida (liverworts) but, instead, pores, which remain permanently open, exist in the surface of the thalli.

Stomata are found in all divisions of the phylum Pteridophyta: in the Psilopsida (e.g., *Psilotum*) large stomata are mainly located in grooves between the stem ridges; the Lycopsida (clubmosses and quillworts) also have large stomata in their leaves and stems; in the Sphenopsida (horsetails and scouring rushes) the stomata are usually confined to the longitudinal grooves of the aerial stems; in the Pteropsida (ferns), stomata are large and abundant in the leaves, usually only on the lower surface.

In the phylum Spermatophyta (higher plants) stomata are most prolific and may be found in the epidermal layers of most aerial parts. They are present in floral parts, including petals, sepals, stamens and gynoecia, in awns, paleas and lemmas of cereal inflorescences, in the inner and outer surfaces of pods and other fruits, including the skins of banana, and have also been recorded in the surface of seeds inside pods. Stomata have also been recorded in rhizomes. More typically they are found in the surface of leaves and green stems and modified leaves such as bracts and tendrils. Stomata are not found in the chlorophyll-free plants, *Monotropa* or *Neottia*, but have been found in the stems of *Orobanche*. Albino plants possess non-functional stomata. Stomata may also be non-functional in some of the locations mentioned above, e.g. petals.

In the leaves of most herbaceous plants stomata are found in both the upper (adaxial) and the lower (abaxial) surfaces but usually there are fewer stomata on the upper surface. Such leaves are termed amphistomatous. Leaves with stomata in the lower leaf surface only, as are usually found in tree species, are called hypostomatous. Aquatic plants with floating leaves, such as water-lilies, have stomata on the upper surface only and such leaves are called epistomatous. In the water fern, *Marsilea* (an aquatic Pteridophyte which can grow on land), stomata are restricted to the upper surface of floating leaves while aerial leaves have stomata on both surfaces. Most submerged aquatic plants do not possess stomata and diffusion of gases occurs between the water and the plant tissues across a thin cuticle.

2.2 Arrangement and frequency of stomata

Usually stomata are randomly scattered within the epidermis. In some species, however, such as *Saxifraga stolonifera* (formerly *sarmentosa*), stomata are located in patches (Figs. 2.5, 2.6). In most monocotyledons and in Gymnosperms stomata are located in rows (Figs. 2.1, 2.2, 2.7, 2.8). Some of the variations in stomatal arrangement between species are illustrated in Figs. 2.1–2.8.

The reason for stomata being roughly equidistant apart in most plants is not completely understood. It has been proposed than an inhibitory zone occurs around a developing stoma preventing formation of other guard mother cells. There may be some hormonal control of stomatal initiation and development. For example, phytochrome may be involved in the development of stomata in mustard cotyledons since in the dark stomata do not fully develop while far-red light results in rapid maturation of stomata and expansion of epidermal cells. Environmental and genetical factors are also involved in the morphogenesis of stomata. Water availability, light intensity, temperature and CO_2 concentration have all been observed to affect stomatal frequency. Stomatal frequencies are usually higher in plants grown in full sunlight than in plants grown in shade. Also, plants growing in a dry soil and low humidity generally have higher stomatal frequencies than plants growing in wet soil and high humidities. However, although stomatal frequency per unit leaf area may vary, since environmental factors may also change leaf areas the number of stomata per leaf may not. For example, Ciha and Brun (1975) found in soybean leaves that increasing temperature caused decreased stomatal frequency per unit leaf area in the abaxial leaf surface, while the frequency increased in both leaf surfaces with increased light intensity. However, because of changing leaf areas brought about by the different light intensities or temperatures, stomata per leaflet did not differ significantly. In the same study, water stress resulted in a greater stomatal frequency, smaller leaf area and significantly lower stomatal numbers per leaflet than in non-stressed leaves.

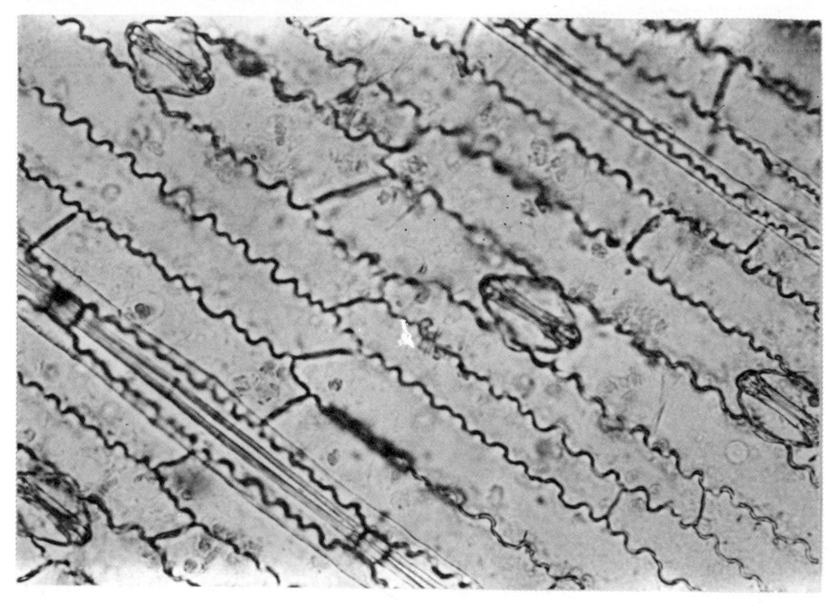

Figure 2.1 Lower leaf surface of maize (*Zea mays*)

Figure 2.2 Scanning electron micrograph of the surface of a needle of Scots pine (*Pinus sylvestris*)

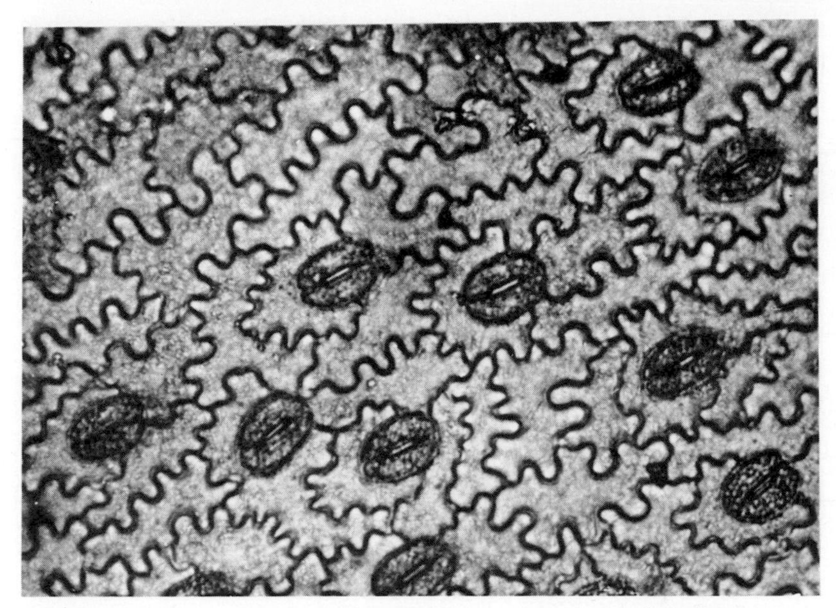

Figure 2.3 Lower surface of a frond of hart's-tongue fern (*Phyllitis scolopendrium*)

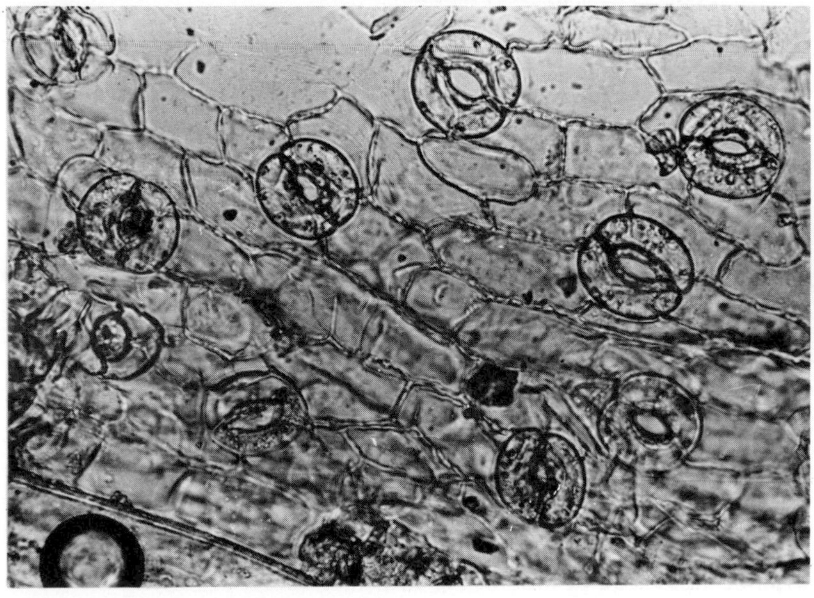

Figure 2.4 Petal of a snapdragon flower

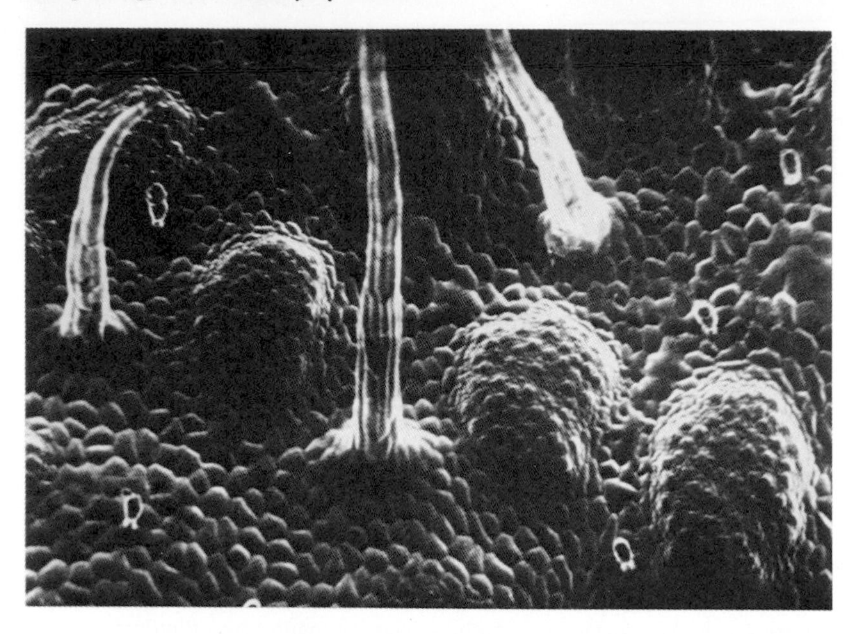

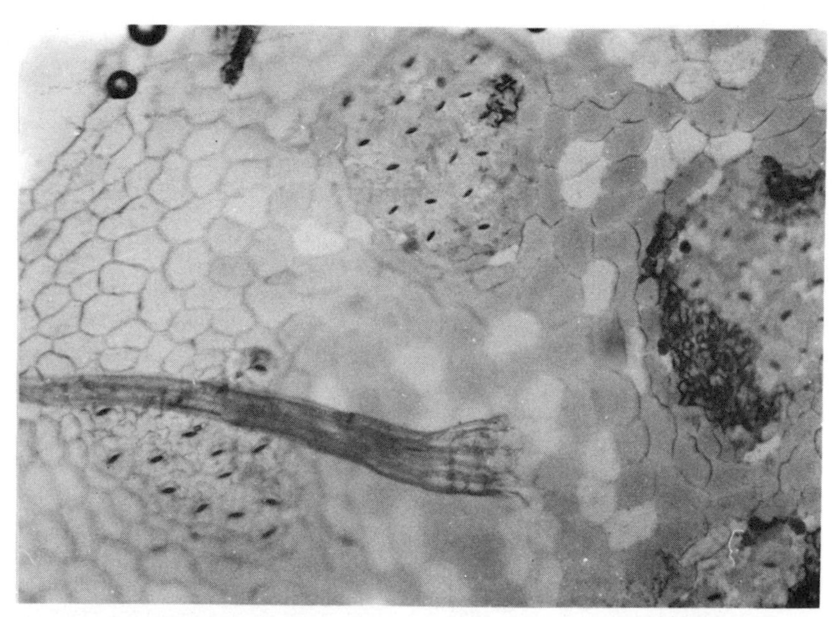

Figures 2.5 and 2.6 Scanning electron micrograph and light micrograph, respectively, of lower leaf surface of *Saxifraga stolonifera* (formerly *S. sarmentosa*) showing the stomata located on raised patches

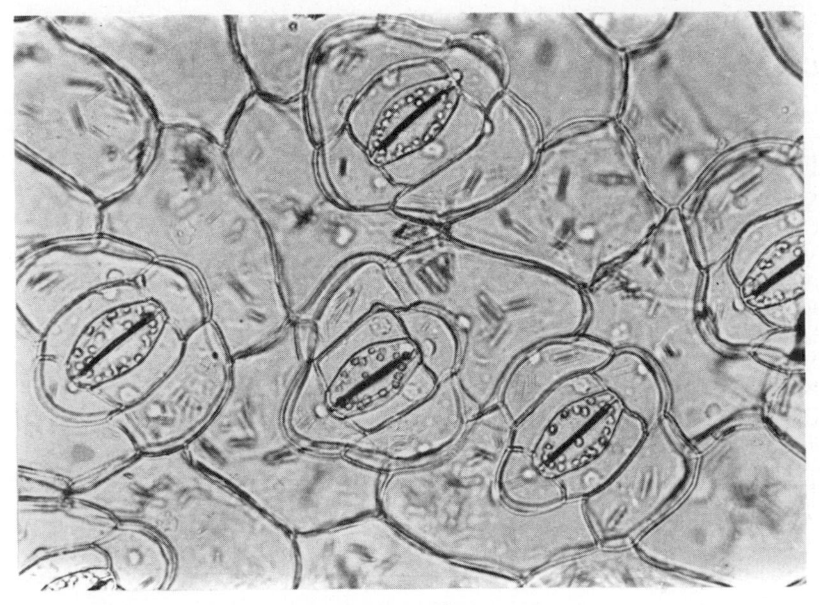

Figure 2.7 Lower leaf surface of *Commelina communis*

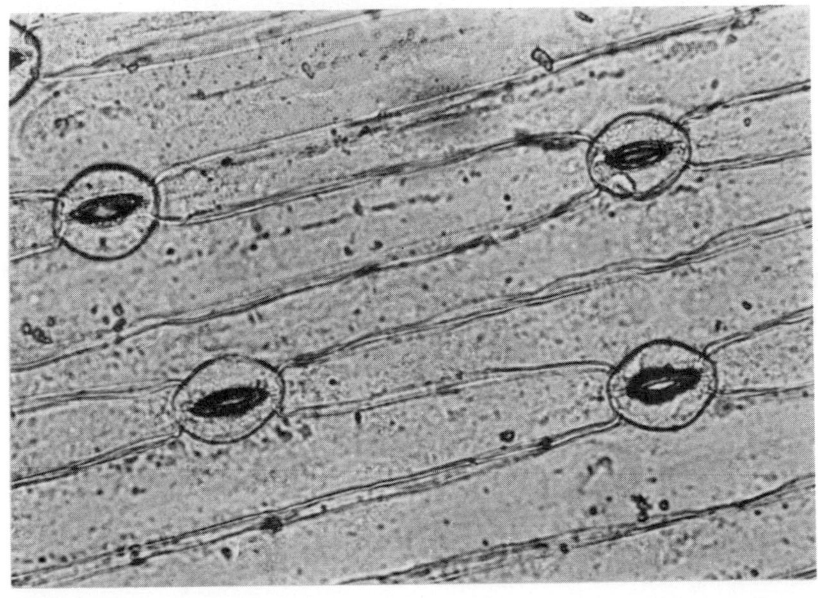

Figure 2.8 Leaf surface of onion (*Allium cepa*)

Table 2.1 Mean stomatal frequency and guard cell length of field grown soybeans (after Ciha and Brun, 1975)

Genotype	**Mean stomatal frequency** (No. of stomata/mm^2)		**Guard cell length** (μm)	
	Adaxial	**Abaxial**	**Adaxial**	**Abaxial**
M-62-93	170	385	22.3	20.8
Corsoy	149	369	22.0	20.0
M-61-93	157	363	21.5	19.5
A. K. Harrow	143	344	21.9	20.9
Blackhawk	130	327	22.5	19.9
Manchu	121	327	22.5	20.2
M-62-56	120	300	21.5	19.6
M-62-263	104	287	21.9	19.4
O. Mandarin	135	263	22.2	20.6
Flambeau	81	242	21.8	20.6

The genotypes have been placed in order of decreasing stomatal frequencies on the abaxial leaf surface. The same trend, though less marked, is observed in the adaxial surface. Stomatal size (in this case represented by guard cell length) usually decreases with increasing frequency but this feature is not observed in the soybean genotypes.

In crop plants stomatal frequency can vary greatly among different genotypes of the same species growing under identical conditions (Table 2.1). Also, as ploidy increases stomatal frequency generally decreases while guard cell dimensions are greater.

Stomatal frequencies also vary within a plant, highest frequencies often being found at the top. Even within a leaf the stomatal frequency may vary; in the grasses, for example, the highest frequency occurs at the point of insertion of the leaf where the cells are smallest and still developing and the lowest frequency may occur at the tip of the leaf.

Because stomatal frequencies often vary according to cell size, smaller guard cells producing highest stomatal frequencies, Salisbury (1928) introduced the term 'stomatal index' which relates the number of stomata per unit leaf area to the number of epidermal cells per unit leaf area. Thus,

$$\text{stomatal index} = \left(\frac{\text{No. of stomata per unit leaf area}}{\text{No. of stomata per unit leaf area} + \text{No. of epidermal cells per unit leaf area}}\right)100$$

and was considered to be fairly constant within the leaves of a single plant. However, this has not been confirmed in all species investigated. In some species the light intensity and quality received by mature leaves of a plant can affect the stomatal index of still developing leaves. For example, in *Vigna sinensis* grown at high irradiance a day of shade causes a decrease in the stomatal index of those leaves that are at a 'critical' period of development (about six days preceding the major phase of unfolding of the leaf lamina) (Schoch *et al.* 1980).

The wide range of stomatal frequencies and guard cell dimensions which occur between species are presented in Table 2.2. Generally, the smaller thc stomata the higher the frequency. This results in the total possible pore area for a leaf being similar for most species. The pore area of a leaf can reach as much as 5 per cent when the stomata are wide open but normally the value remains below 2 per cent. Tree species generally have high stomatal frequencies while xerophytes generally have low frequencies. Why there are such large variations in stomatal frequency and guard cell dimensions and what is the functional significance of these variations are not fully understood.

2.3 Anatomy and morphology of the epidermis of leaves

The epidermis prevents excessive and uncontrollable water loss by the leaf, reduces infection of the leaf by fungal spores and bacteria, and minimises mechanical damage to the leaf.

The morphology and anatomy of the epidermal layer varies greatly between species and the epidermis may contain a variety of cell types, including trichomes, epidermal cells (including cork cells, silica cells, bulliform cells and long cells, all of which are found in grass species), subsidiary (accessory) cells and guard cells. The various features and dimensions of cells of the epidermes of a maize leaf and their relationship to other cells within the leaf area are shown in Fig. 2.9A and B.

2.3a Cuticle

The epidermal layer is covered by a layer of cuticle which varies in thickness from species to species and within a species depending on the growth conditions of the plants. Although there is some disagreement about the structure and chemistry of the cuticle it essentially consists of a matrix of cutin (which may consist of large polymers of hydroxylated C_{16} and C_{18} fatty acids), cellulose, oligosaccharides, and wax. The cuticle has been split up into zones, each zone being slightly different from the other. The lower zone impregnates the outer cellulose cell wall layers while wax of varying thickness is deposited on the outer surface of the cuticle.

Humid conditions may cause the cuticle to swell and become more permeable to water. Water may also be excreted through microscopic pores, particularly above hydathodes, by a process known as guttation.

It is now considered that a cuticular layer completely surrounds cell wall surfaces in direct contact with air. Thus, the cuticle extends from

Table 2.2 Stomatal frequencies, guard cell dimensions and pore area as a percentage of leaf area when stomatal apertures are 6 μm

Species	Stomatal frequency (per mm^2)*		Pore Area (% of leaf area with apertures of 6 μm)	Guard cell dimensions for lower epidermis* (μm)		Ref.
	Lower epidermis	Upper epidermis		length	width	
FERNS						
Phylilis scolopendrium (hart's tongue-fern)	59	0	0.55	77	21	Meidner and Mansfield (1968)
Osmunda regalis (Royal fern)	67	0	0.50	56	31	Meidner and Mansfield (1968)
HERBACEOUS PLANTS						
Monocots						
Commelina communis	67	19	—	48	12	Willmer, unpub.
Zea mays (maize)	108	98	0.70	43	12	Meidner and Mansfield (1968)
Triticum vulgare (wheat)	40	50	0.63	53	14	Meidner and Mansfield (1968)
Hordeum vulgare (barley)	85	70	0.65	38	11	Meidner and Mansfield (1968)
Avena sativa (oat)	45	50	0.50	56	13	Meidner and Mansfield (1968)
Allium cepa (onion)	175	175	0.63	24	12	Meidner and Mansfield (1968)
Paphiopedilium venustum (Lady slipper orchid)	26	0	—	67	53	Rutter and Willmer (1979)
Tradescantia virginiana	23	7	0.35	70	21	Meidner and Mansfield (1968)
Dicots						
Helianthus annuus (sunflower)	175	120	1.10	32	15	Meidner and Mansfield (1968)
Vicia faba (broad bean)	75	65	1.00	46	13	Meidner and Mansfield (1968)
Sedum spectablis	35	28	0.32	33	11	Meidner and Mansfield (1968)
Nicotiana tabacum (tobacco)	190	50	0.80	31	13	Meidner and Mansfield (1968)
Xanthium pennsylvanicum	173	177	0	39	12	Meidner and Mansfield (1968)

SMALL TREES AND SHRUBS						
Corylus americana (hazelnut)	347	0	—	37	—	
Cornus florida (flowering dogwood)	83	0	—	32	—	
Rhus copallina (shining sumac)	731	0	—	6	—	Carpenter and Smith (1975)
LARGE TREES						
Angiosperm						
Quercus palustris (pin oak)	909	0	—	10	—	
Quercus valutina (black oak)	405	0	—	50	—	
Quercus triloba (Spanish oak)	1,192	0	0.43	—	—	Spector (1956)
Tilia americana (American basswood)	891	0	—	26	—	Carpenter and Smith (1975)
Tilia europa (European lime)	370	0	0.90	12	9	
Gymnosperms						
Pinus sylvestris (Scots pine)	120	120	1.20	28	14	Meidner and Mansfield (1968)
Larix decidua (larch)	16	14	0.15	42	13	
OTHER						
Welwitschia mirabilis (a gymnosperm with CAM)	222	222	—	—	—	Bornman (1972)

* Stomatal frequencies and dimensions depend on a variety of factors including their location on the leaf, the leaf insertion and growth conditions of the plants and the values given here should not be accepted as definitive ones.

— no figures available.

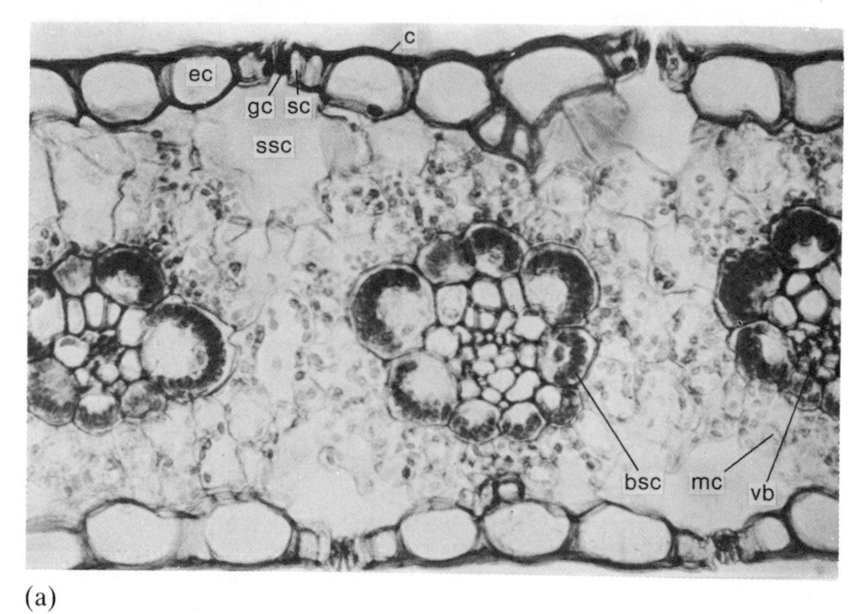

(a)

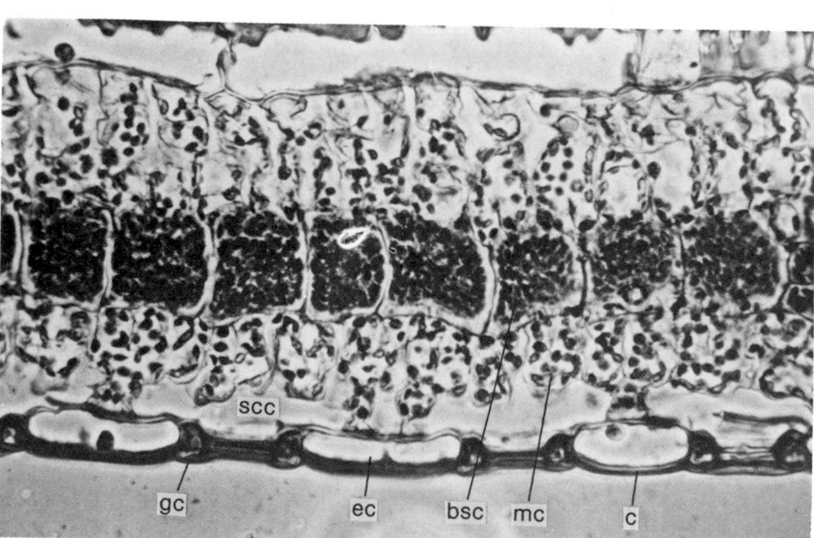

(b)

Figure 2.9 (a) Transverse section and (b) longitudinal section of a *Zea mays* leaf showing the typical C_4 anatomy and the typical arrangement of stomata in the Graminae. Note, also, how small guard cells are in relation to the size of other cell-types within a leaf. gc = guard cell, sc = subsidiary cell, ec = epidermal cell, mc = mesophyll cell, bsc = bundle sheath cell, vb = vascular bundle, c = cuticle, ssc = substomatal cavity.

the leaf surface through stomatal pores and then, as a very thin layer, covers walls exposed to air in the substomatal cavity and other leaf air spaces. Indeed, if cuticle did not cover cell walls exposed to leaf air spaces, under certain conditions it has been suggested that the air spaces would fill up with water.

2.3b Trichomes

Trichomes show great variation in form and structure. They may be unicellular, or multicellular and secretory or non-secretory (*see* Esau 1965). Their purpose may be to excrete unwanted materials from leaves, such as NaCl which may accumulate in halophytes, or to secrete aromatic compounds either to attract pollinators or distract herbivores. Trichomes also decrease water loss from leaves by increasing the boundary layer resistance.

In certain metabolic aspects trichomes may resemble guard cells; they are often highly active cells involved in secretory processes and they have a dense cytoplasmic content with many mitochondria. Also, the cells comprising the trichomes may possess high concentrations of K^+. Thus, it is advisable that epidermes used for studies of stomata should be taken from species lacking trichomes.

2.3c Epidermal cells

Usually epidermal cells are the largest cell type within the epidermal layer and they show considerable variation in size and shape. In the grasses special epidermal cells called cork cells and silica cells occur which are relatively small, and have dense cytoplasmic contents. Usually, however, epidermal cells are highly vacuolated and the peripheral layer of cytoplasm contains a low frequency of cell organelles. Although normally absent, chloroplasts may be present in epidermal cells to varying extents; in *Vicia faba* they are few in number while in many ferns they are very abundant. The vacuoles of epidermal cells may often contain large numbers of inclusions, such as the calcium oxalate crystals found in *Commelina communis*, anthocyanins, flavonoids and alkaloids.

The functions of the epidermal cells are many and varied. They may act as depositories for unwanted material such as calcium or oxalate or as a storage site for water. The large watery vacuoles also play an important role in absorbing harmful UV radiation and in absorbing infra red radiation to dissipate the heat. Epidermal cells have also been implicated in the perception of stimuli. The thicker, cuticularised outer walls of the epidermal cells also resist attack by insects and other pests and pathogens as well as preventing excessive water loss. Those epidermal cells containing chloroplasts may also be photosynthetic. Bulliform cells are a type of epidermal cell prevalent in grasses. The cells are large, thin-walled and highly vacuolated. Their exact function is not certain: they may be involved in unrolling developing leaves, or in the opening and closing movements of mature leaves or in water storage.

2.3d Subsidiary cells

Subsidiary cells (also known as accessory cells) are specialised epidermal cells which are usually different in size and shape from other epidermal cells and lie adjacent to, or in close proximity (in some cases where more than one subsidiary cell exists) to, the guard cells. In some species subsidiary cells lie above the guard cells as in pine (Fig. 2.4) and *Equisetum* (Fig. 2.2). Subsidiary cells are more complex in terms of structure and function than epidermal cells. They are usually smaller than epidermal cells and contain a more dense cytoplasmic content and a greater frequency of cell organelles but they do not normally contain chloroplasts, anthocyanins or crystalline inclusions. In the grasses and some other species subsidiary cells are intimately involved in the stomatal mechanism.

2.3e Guard cells

Guard cells are not only the most complicated cells biochemically and anatomically within the epidermal layer but are also unique amongst the cells of the plant body. They possess a specialised set of metabolic pathways designed to bring about rapid changes in osmotic potential within the cells when stimulated by relatively minor changes in the external environment. For example, a change in the CO_2 concentration around guard cells of only a few dozen $\mu l\ l^{-1}$ may set in motion a series of reactions resulting in stomatal opening or closing. Associated with the stomatal movements are large and rapid fluxes of K^+ and H^+, and usually, to a lesser extent Cl^-, between the guard and neighbouring cells. Guard cells have a high metabolic activity relative to most other cell types and they contain an abundance of mitochondria (Fig. 3.13) and protein synthesising machinery (Fig. 2.10).

Although guard cells are relatively small cells there is considerable variation in their size. Depending on the species the dimensions of the outer limits of the walls may vary from under 10 μm to almost 80 μm in length and from a few μm to about 50 μm in width (Table 2.2), though the dimensions of width may vary according to the stomatal aperture (*see* Ch. 4). Table 2.3 gives some other dimensions and characteristics of guard cells and compares them with values from palisade and spongy mesophyll cells of the same species.

From an anatomical viewpoint there are two basic types of stomata though intermediates may exist in gymnosperms and sedges

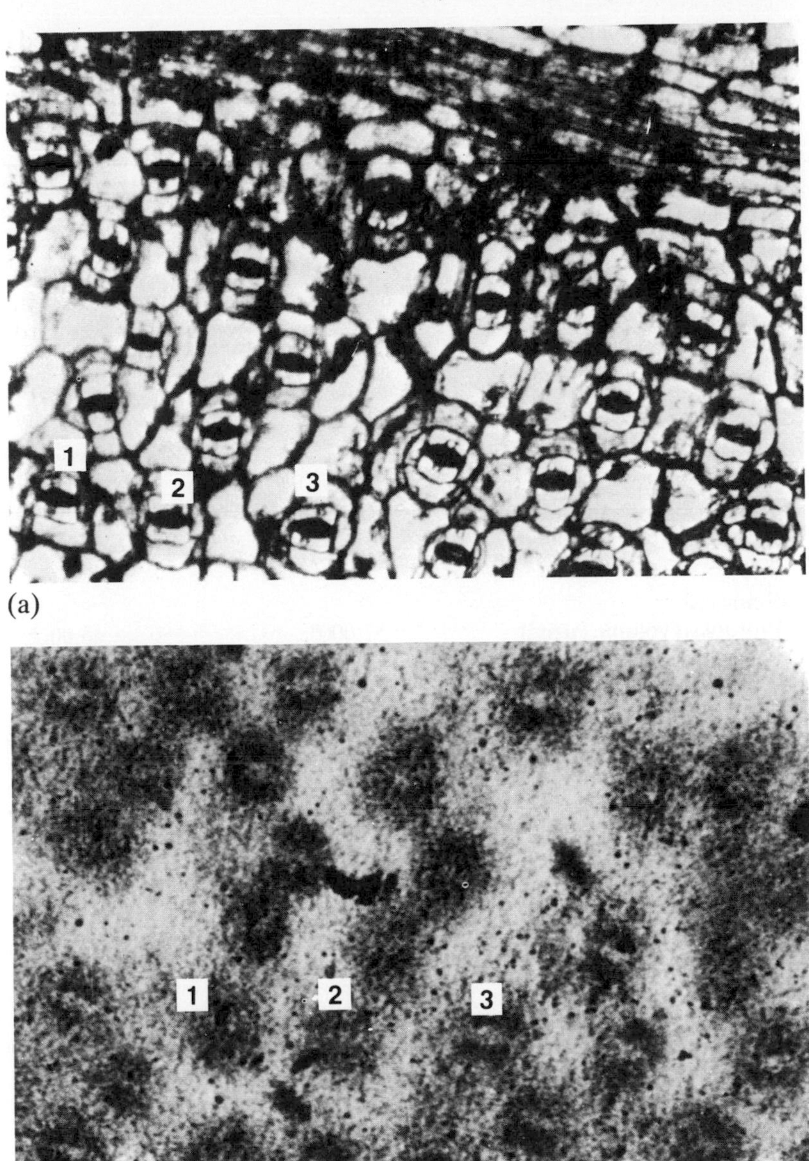

Figure 2.10 (a) freeze-dried epidermis of *Commelina communis* which has been exposed to ^{3}H-leucine; (b) autoradiograph of the tissue after extraction of all ethanol soluble material. The blackened areas which appear over the stomata (stomata 1, 2, 3 correspond to blackened areas 1, 2, 3) represent location of high protein synthesis (Willmer, Bernal-Lugo and Varner, unpublished).

Table 2.3 Some characteristics of guard cells, palisade cells and spongy mesophyll cells. [a], Pearson and Milthorpe (1974); [b], Outlaw and Lowry (1977); [c], Outlaw *et al.* (1980); [d], Jones *et al.* (1977); [e], Outlaw *et al.* (1981); [f], Outlaw (private communication)

Cell type	*Commelina cyanea*	*Vicia faba*
Guard cell		
Protoplast volume *(μm^3)		
stoma open to 10 μm	7,200[a]	2,800[a]
stoma closed	3,200[a]	1,500[a]
Dry weight (ng)		
stoma open to 10 μm	—	3.4[b]
stoma closed	—	2.9[b]
Chloroplasts/cell	10[a]	8–10[a,c]
Chloroplast volume (μm^3)	17[a]	12[a]
Chlorophyll (pg)/chloroplast	—	0.3–0.5[c]
Soluble protein/cell (% of dry weight; average of open and closed stomata)	—	5.2[e]
Soluble protein : chlorophyll	—	35 : 1[c]
Palisade cell		
Protoplast volume (μm^3)	57,000[a]	46,000[a]
Dry weight (ng)	—	11.6[d]
Chloroplasts/cell	41[a]	59[a]
Chloroplast volume (μm^3)	64[a]	33[a]
Chlorophyll (pg)/chloroplast	—	2.1[c]
Soluble protein/cell (% of dry weight)	—	21.1[e]
Soluble protein : chlorophyll	—	20.1[f]
Spongy mesophyll cell		
Protoplast volume (μm^3)	48,000[a]	42,000[a]
Dry weight (ng)	—	13.8[d]
Chloroplasts/cell	28[a]	24[a]
Soluble protein/cell (% of dry weight)	—	17.6[e]
Soluble protein : chlorophyll	—	40.1[f]

* 1 μm^3 = 1 nl
— = No figures available

(Cyperaceae). One type has kidney-shaped guard cells and elliptical stomata (Fig. 2.11A, B), while the other type, known as the grass-type, has dumb-bell-shaped guard cells (Fig. 2.11C, D). The dumb-bell-shaped guard cells are restricted to the monocotyledons and, although found in all Gramineae, are also reported to be in the Cyperaceae, Lepidocaryoid and Arecoid Palmae, Flagellariaceae, Rapateaceae, Marantaceae, Anarthriaceae and Loweriaceae.

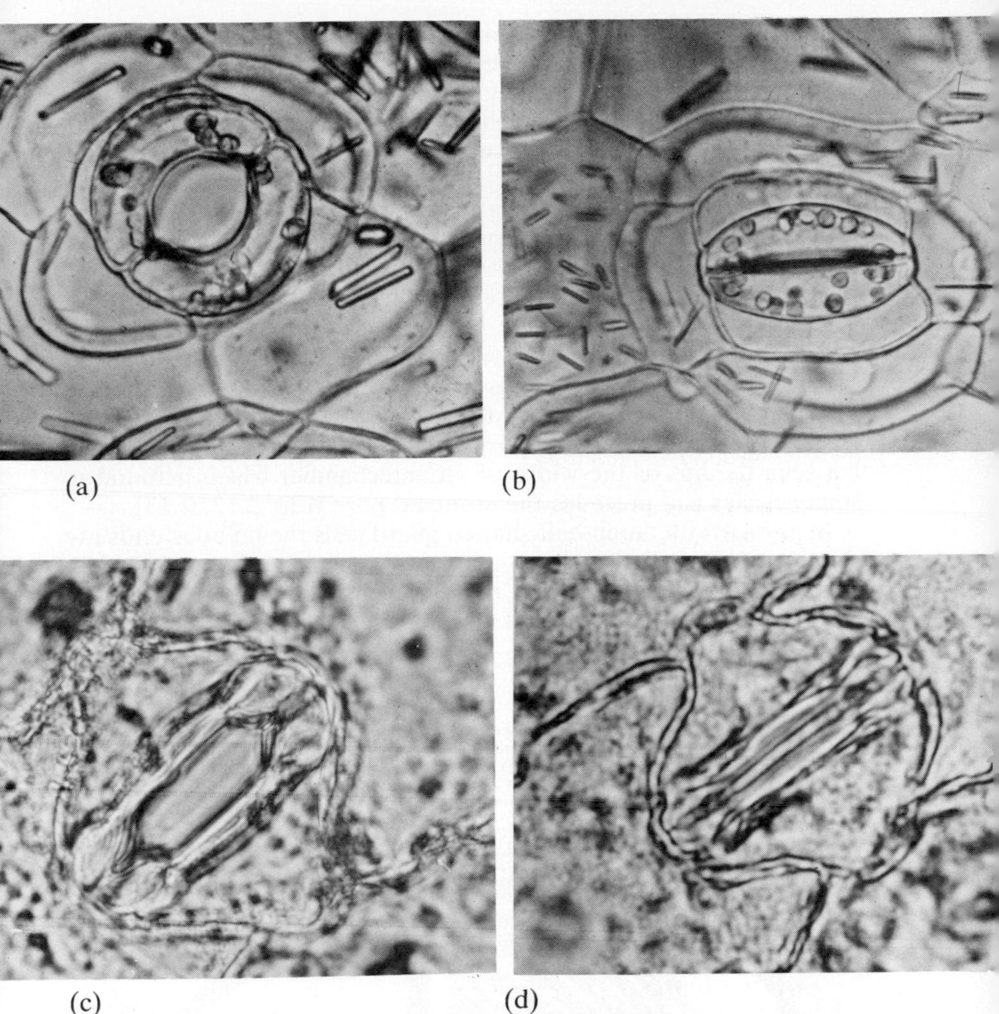

Figure 2.11 (a) open and (b) closed stoma of *Commelina communis* which has kidney- or sausage-shaped guard cells and elliptical stomata, (c) open and (d) closed stoma of *Zea mays* which has dumb-bell-shaped guard cells and 'grass-type' stomata. Note also that *C. communis* has two lateral pairs of subsidiary cells and a terminal pair. When stomata open the inner lateral pair become squashed by the guard cells. The long, rectangular crystals in all but the guard cells and inner lateral subsidiary cells are of calcium oxalate. Notable features in *Z. mays* are the pair of triangular-shaped subsidiary cells and the ventral walls of each guard cell which remain parallel to each other as the stoma opens.

Conventionally, plant anatomists have called the wall facing the pore of the guard cell the ventral wall while the opposite wall juxtaposed to the neighbouring cell is termed the dorsal wall. The walls parallel to the surface of the leaf facing the atmosphere and the substomatal cavity are known as the outer and inner lateral walls, respectively.

Guard cell wall thickening varies greatly according to species and location around the cell. In kidney-shaped guard cells the dorsal wall is usually thin while the ventral wall is usually heavily thickened and the surface facing the pore may be sculptured in a variety of ways. The pore is usually guarded by a pair of lips (usually heavily cuticularised) protruding from the upper edge of the ventral wall. In some species a lower pair of lips may also exist, while, in some other species, lips may be entirely absent. The term stomatal aperture refers to the width of the pore at the throat and not the width between the lips or the width of an antechamber which is found in some species and precedes the stomatal pore (Fig. 2.12, 2.13).

In grasses with dumb-bell-shaped guard cells the bulbous ends are relatively thin-walled and are connected to each other by a thick-walled middle section. In *Zea mays* and probably other grasses the middle portion has thin ventral and dorsal walls but very thick upper and lower lateral walls.

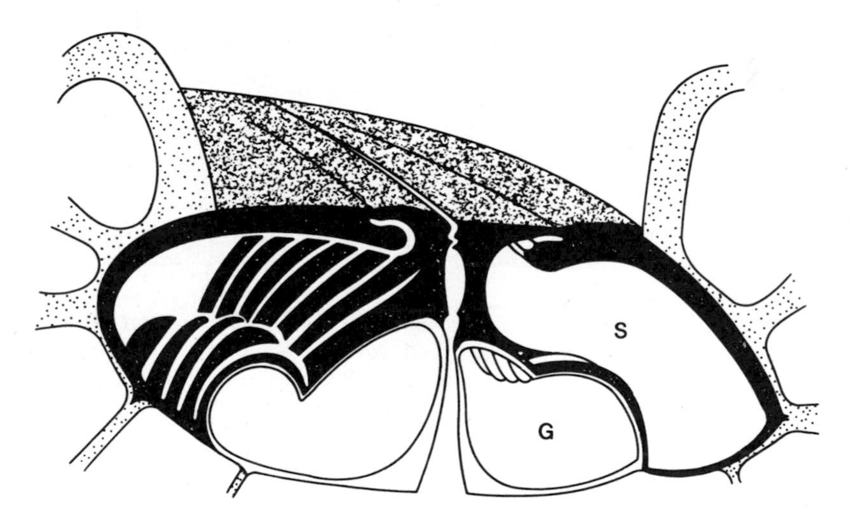

Figure 2.12 Diagrammatic representation of a stomatal complex of *Equisetum* spp in the subgenus *Hippochaete*. G = guard cell; S = subsidiary cell. Note the ridges on the common wall between the subsidiary and guard cells and the interlocking ridges at the entrance of the antechamber preceding the stomatal pore (after Dayanandan and Kaufman 1973).

Figure 2.13 A stoma of *Paphiopedilum insigne* (hybrid) under Nomarski interference contrast illumination. Note the prominent cuticularised lips extending over a large antechamber, the thick cuticle and the heavily thickened guard cell walls, particularly the inner lateral ones (from Rutter and Willmer 1979).

Some stomata and stomatal complexes are very elaborately structured. In *Equisetum* conspicuous ridges exist on the joint subsidiary cell/guard cell wall and numerous infoldings and inter-locking ridges occur in the subsidiary cell wall (Fig. 2.1). In *Pinus* the wall thickenings of the stomatal complex are also unusual in that there are regions of lignified material in the upper and lower lateral guard cell walls with a very thin wall area facing the pore and heavily lignified subsidiary cell walls (Fig. 2.14). Normally, guard cells are attached to other cells by their dorsal wall only but in some genera such as *Pinus* and *Equisetum* a number of cells, or just one subsidiary cell, envelop the guard cell, making a relatively rigid structure (Figs. 2.12 and 2.14). In such cases, as the stomata develop and lignification increases, they may become non-functional.

The nature of the guard cell wall is not completely understood. It appears to be rich in pectins which impregnate the cellulose

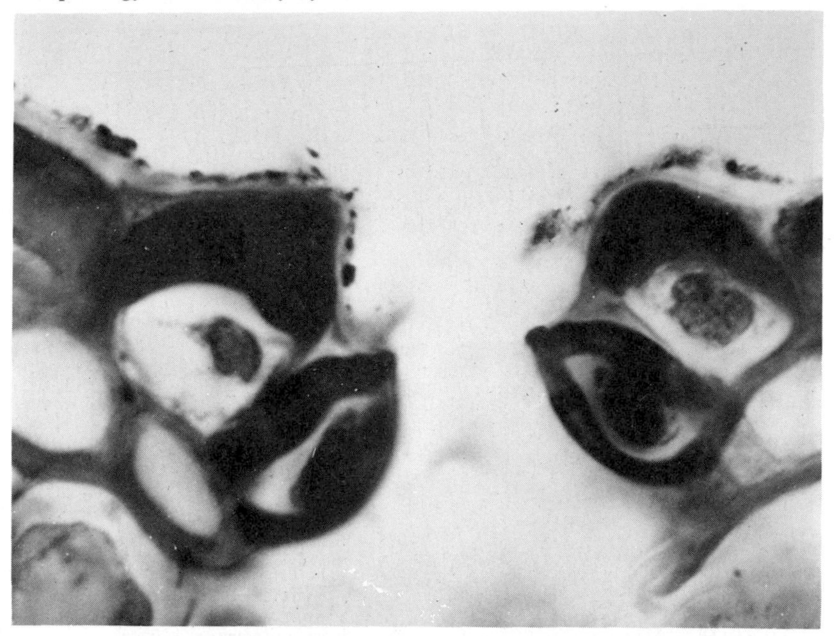

Figure 2.14 A stomatal complex of Scots pine (*Pinus sylvestris*) showing a sunken stoma and the lignified areas (darkened regions) in the guard and subsidiary cell walls.

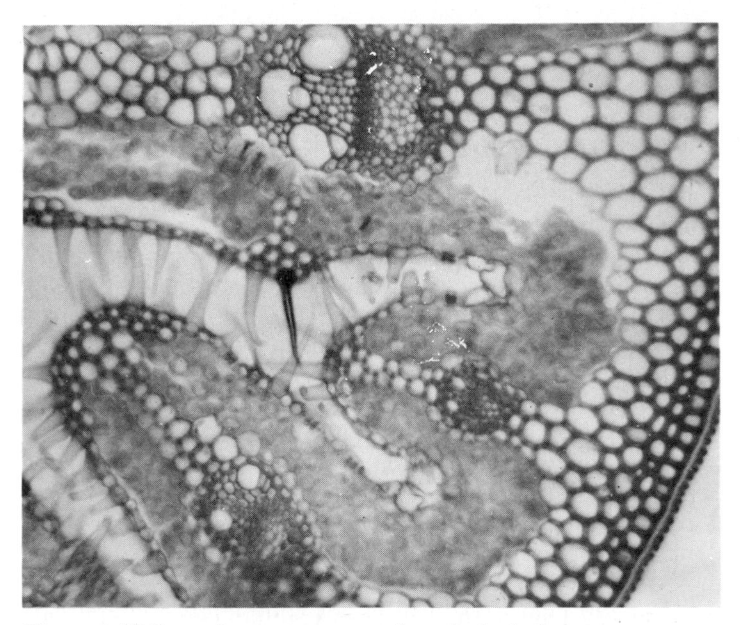

Figure 2.15 Part of a transverse section of a leaf of marram grass (*Ammophilia arenaria*) showing deep grooves, at the bottom of which are located the stomata.

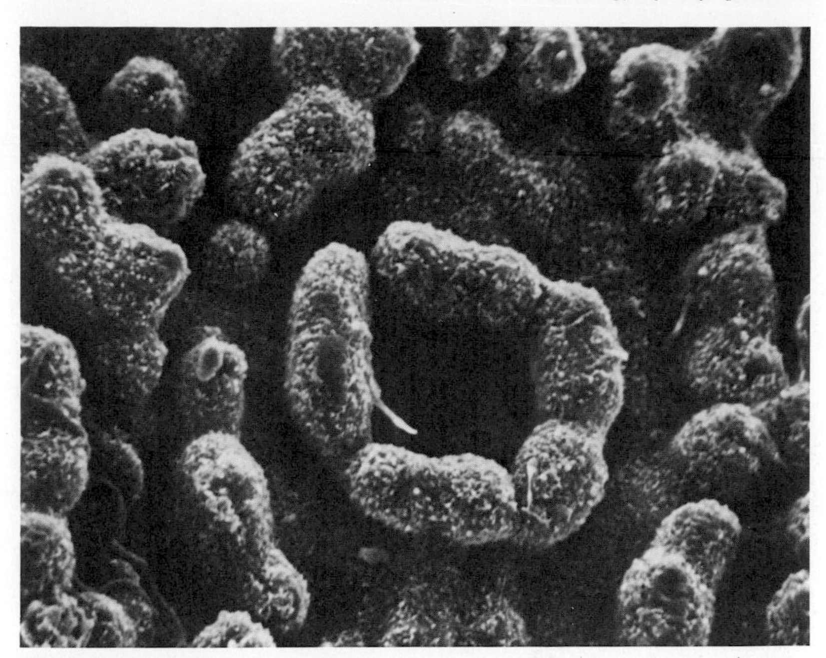

Figure 2.16 Scanning electron micrograph of yew (*Taxus baccata*) showing a sunken stoma and large epidermal cell protuberances which will increase the boundary layer resistance of the leaf.

microfibrils. Microscopy with polarised light (Ziegenspeck 1938) and electron microscopy indicate a predominantly radial orientation of the cellulose microfibrils (micellae) in the kidney-shaped guard cells, i.e. radiating out from the pore around the guard cell in the manner found in a radial tyre (*see* §4.5 and Fig. 4.5). In the grasses the micellae radiate out from the pore at the bulbous ends of the guard cells with an axial arrangement predominating in the connecting pieces between the bulbous ends (*see* §4.5 and Fig. 3.7). The guard cell walls of gymnosperms and some ferns contain lignin while a β-1,3 glucan (probably callose) has been located next to the plasmalemma in the guard cell walls of species of *Ophioglossum* (a fern). Deposits of silicon (probably existing as silicon dioxide) have also been detected in the middle lamellar region and just below the cuticle of guard cells and subsidiary cells of sugar cane and in the walls of guard cells and subsidiary cells of *Equisetum* and some grass species.

Stomata may be located in pits as in marram grass (Fig. 2.15) or sunken below the surface of the leaf as in conifers (Figs. 2.2, 2.14). Also, in the conifers the pits leading to the stomatal pores gradually fill up with intermeshed wax tubules. Often sunken stomata have subsidiary and/or epidermal cells growing above the guard cells as in *Equisetum* and gymnosperms (Figs. 2.12, 2.14) and the opening to the leaf surface may be surrounded by a complex surface sculpturing of wax crystals or wall protruberances (e.g., Fig. 2.16). All these features assist in decreasing water loss from plants by increasing the boundary layer resistance.

In some species, e.g. *Saxifraga stolonifera* (formerly *sarmentosa*), stomata project above the level of the leaf surface (Figs. 2.5, 2.6) which would assist in increasing transpirational water loss. Stomata may also be able to move in the leaf surface as though in sockets. 'Hautgelenke' or hinges located at the top and bottom of the dorsal wall of the guard cell have been reported (Schwendener 1881) which are purported to facilitate such guard cell movements.

Further reading and references

Allaway, W. G. and Milthorpe, F. L. (1976) Structure and functioning of stomata, in *Water Deficits and Plant Growth*, vol. IV. Academic Press, NY.

Bornman, C. H. (1972) *Welwitschia mirabilis*: Paradox of the namib desert, *Endeavour*, **31**, 95–9.

Carpenter, S. B. and Smith, N. D. (1975) Stomatal distribution and size in southern Appalachian hardwoods, *Can. J. Bot.*, **53**, 1153–6.

Ciha, A. J. and Brun, W. A. (1975) Stomatal size and frequency in soybeans, *Crop Sci.*, **15**, 309–13.

Copeland, E. B. (1902) The mechanism of stomata, *Ann. Bot.*, **16**, 327–64

Dayanandan, P. and Kaufman, P. B. (1973) Stomata in *Equisetum, Can. J. Bot.*, **51**, 1555–64.

Esau, K. (1965) *Plant Anatomy* (2nd edn). Wiley, NY.

Jones, M. G. K., Outlaw, W. H. Jr, and Lowry, O. H. (1977) Enzymic assays of 10^{-7} to 10^{-14} moles of sucrose in plant tissues, *Plant Physiol.*, **60**, 379–83.

Meidner, H. and Mansfield, T. A. (1968) *Physiology of Stomata.* McGraw-Hill, London.

Outlaw, W. H. Jr, and Lowry, O. H. (1977) Organic acid and potassium accumulation in guard cells during stomatal opening, *Natl. Acad. Sci., USA*, **74**, 4434–8.

Outlaw, W. H. Jr, Manchester, J. and Zenger, V. E. (1981) The relationship between protein content and dry weight of guard cells and other single cell samples of *Vicia faba* L. leaflet, *Histochem. J.*, **13**, 329–36.

Outlaw, W. H. Jr, Mayne, B. C. Zenger, V. E. and Manchester, J. (1980) Presence of both photosystems in guard cells of *Vicia faba* L: Implications for environmental signal processing, *Plant Physiol.*, **67**, 12–16.

Pearson, C. J. and Milthorpe, F. L. (1974) Structure, carbon dioxide fixation and metabolism of stomata, *Aust. J. Plant Physiol.*, **1**, 221–36.

Rutter, J. C. and Willmer, C. M. (1979) A light and electron microscopy study of the epidermis of *Paphiopedilum* spp. with emphasis on stomatal ultrastructure, *Plant Cell Environment*, **2**, 211–19.
Salisbury, E. J. (1928) On the causes and ecological significance of stomatal frequency with special reference to woodland flora, *Phil. Trans. R. Soc. B.*, **216**, 1–65.
Schoch, P.-G., Zinsou, C. and Sibi, M. (1980) Dependence of the stomatal index on environmental factors during stomatal differentiation in leaves of *Vigna sinensis* L., *J. Exp. Bot.*, **31**, 1211–16.
Schwendener, S. (1881) Über Bau und Mechanik der Spaltöffnungen, *Monatsber. Kgl. Akad. Wiss. Berlin*, **43**, 833–67.
Spector, (1956) In *Handbook of Biological Data.* XXXVI, p. 146. Sanders Co., Philadelphia, USA.
Von Guttenberg, H. (1971) Bewegungsgewebe und Perzeptionsorgane, in *Handbuch der Pflanzenanatomie.* Carlquist, S., Zimmermann, W. Ozenda, P., Wulff, H. D. (eds) Gebrüder Borntraeger, Berlin, Stuttgart. Vol. V 5, 203–19
Ziegenspeck, H., (1938) Die Micellierung der Turgeszen-mechanismen. Tiel 1. Die Spaltöffnungen (mit phylogenetischen Ausblicken), *Bot. Arch.*, **39**, 268–309, 332–372.

3 Stomatal development and differentiation and the ultrastructure of guard cells

3.1 Development and differentiation of stomata

The development and differentiation of stomata and stomatal complexes has particularly attracted the attention of cell biologists because certain basic cellular processes such as cell determination, intracellular mobility and functional divergence are often easily observed. Additionally, leaves from monocotyledons, particularly the grasses, have been very useful in studying stomatal differentiation because of the orderly sequence of events which occurs from the base of the leaf, where the youngest tissue is located, towards the tip (centripetal development).

Guard cells arise from protodermal cells. A protodermal cell becomes polarised, i.e. one end of the cell – the distal end in *Allium cepa* – accumulates most of the cytoplasm and organelles, including the nucleus, and then divides mitotically, giving rise to two cells of unequal size (Bünning and Biegert 1953). The smaller distal cell with dense cytoplasmic contents becomes the guard mother cell (sometimes called the guard cell mother cell) while the larger proximal cell becomes a less specialised cell. The guard mother cell then divides symmetrically and longitudinally to form the guard cells (Figs 3.1 to 3.5).

Although the fine details of the division of the guard mother cell giving rise to elliptical stomata may differ between species the basic features should be common to all species. An excellent study made by Palevitz and Hepler (1974a, b) of the development of elliptical stomata of onion, where the process is not complicated by the formation of subsidiary cells, may serve as an example.

At interphase the guard mother cells are usually rectangular in shape and about 9 × 14 μm in size. They contain several small vacuoles and all the usual cell organelles including lipid bodies, microbodies and proplastids which, in most species, contain starch at this early stage of development. Dictyosomes and polysomes are abundant and microtubules are distributed around the cell adjacent

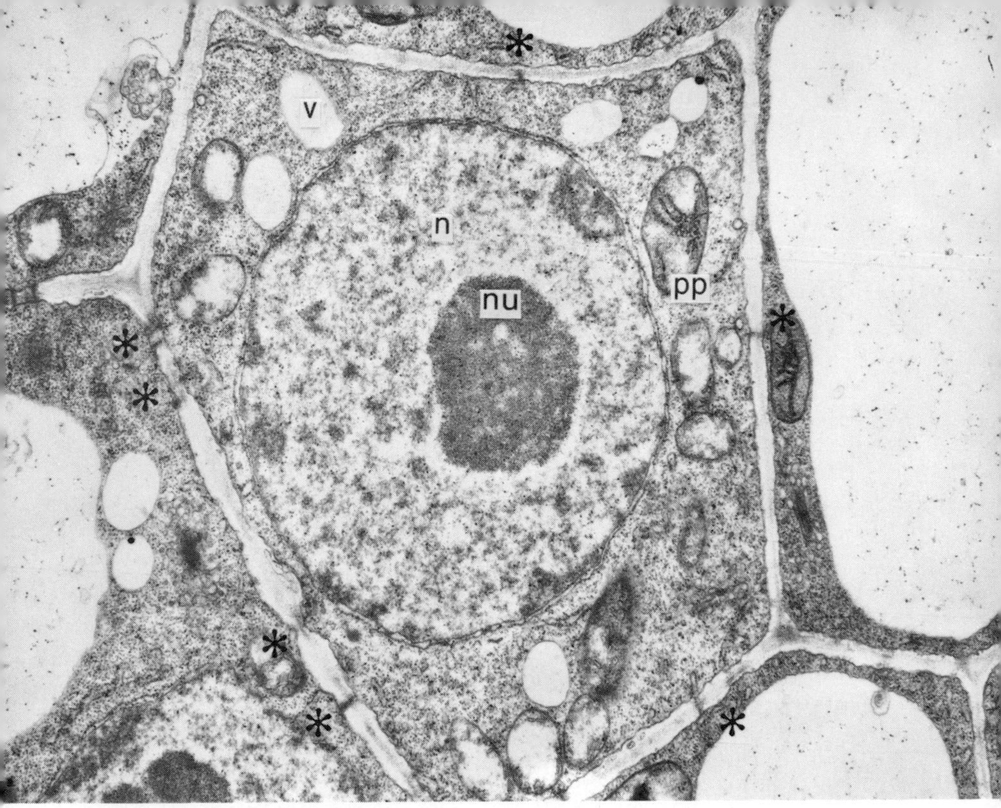

Figure 3.1 (× 15,000) Guard mother cell of *Phaseolus vulgaris* with proplastids (pp) containing starch, a few small vacuoles (v), a central nucleus (n) with nucleolus (nu). Plasmodesmata (*) join the guard mother cell with neighbouring epidermal cells which are vacuolated.

to the walls. At late interphase/pre-prophase a band of microtubules becomes evident at the centre of the proximal and distal ends of the guard mother cell stretching across from the nucleus while the other peripheral layer of microtubules disappears.

At prophase the band of microtubules becomes more evident with a layer of them now surrounding the nucleus. The chromosomes continue to condense, the nucleoli(us) having broken down by this time, and the nuclear envelope begins to disintegrate.

During metaphase the chromosomes become orientated on the metaphase plate (congression) with their arms extended away from the plate. The plate usually forms at an oblique angle to the long axis of the cell, often with its edges pointing to diametrically opposite corners. This feature has been observed in other very small dividing cells and it has, therefore, been suggested that the tilting of the plate is due to the limited space within the cell.

At anaphase the daughter chromosomes separate to opposite corners of the guard mother cell. The separation may also involve

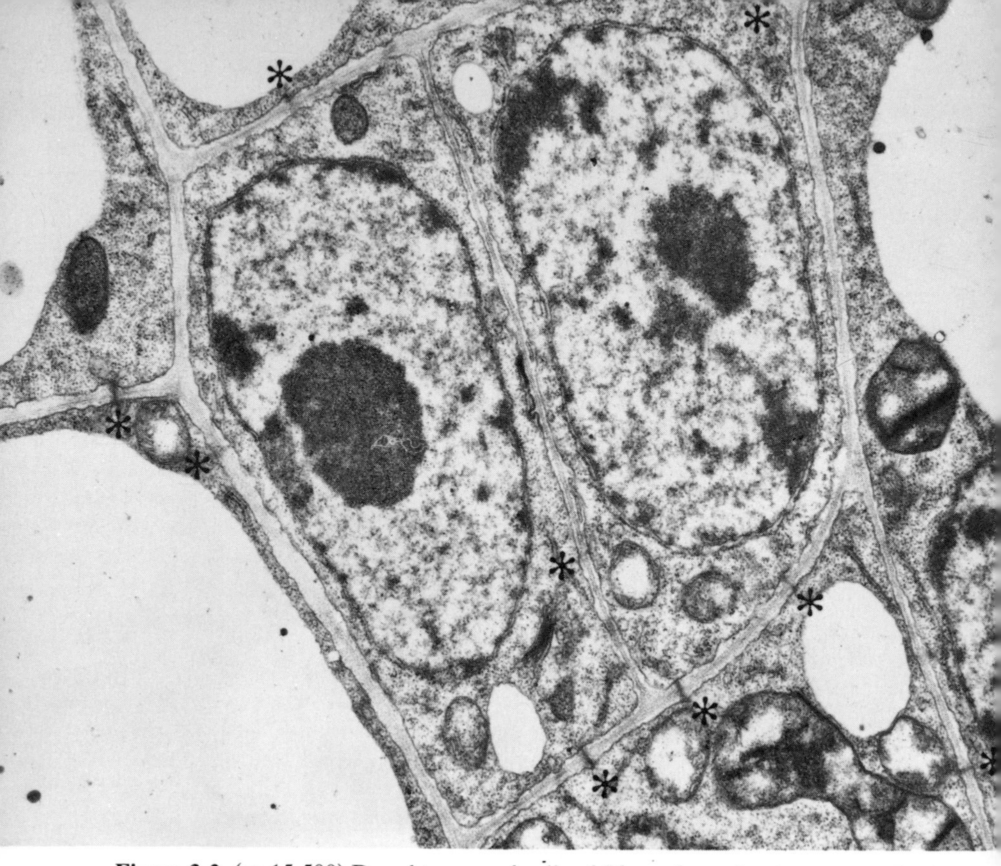

Figure 3.2 (× 15,500) Daughter guard cells of *Phaseolus vulgaris* connected to each other by plasmodesmata (*)

one nuclear mass moving to the outer lateral surface of the cell while the other nuclear mass moves towards the inner lateral surface. Thus, the nuclei come to be diagonally opposite from the top corner to the bottom corner of the cell.

At later anaphase/early telophase large numbers of refractile particles accumulate near the centre of the interzone region next to the developing cell plate. The developing plate may appear as a single structure or as two shorter separate pieces which fuse before growing centrifugally towards the corners. The plate is formed from the fusion of small vesicles derived from dictyosomes or endoplasmic reticulum. The phragmoplast microtubules appear to align themselves in the direction of the daughter nuclei and even contact them. The reorientation of the cell plate is also initiated at this stage and eventually the entire spindle apparatus, including daughter nuclei and growing phragmoplast, rotate from the oblique position to one in which the cell plate is orientated parallel to the long axis of the dividing guard mother cell. The nuclei are then opposite each other in the same

84-4400 MCL

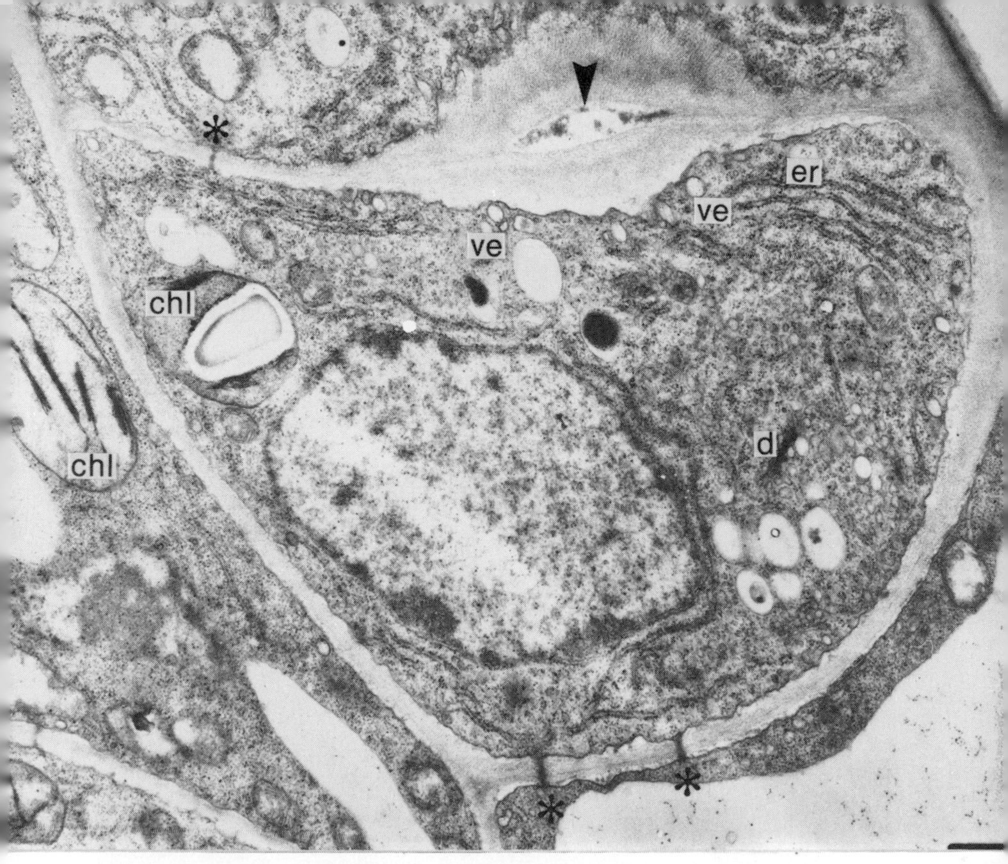

Figure 3.3 (Mag. × 12,500) developing stoma of *Phaseolus vulgaris* with pore (←) partly formed. Plasmodesmata (*) occur between sister guard cells in the ventral walls and between guard cells and epidermal cells. Rough ER (er), dictyosomes (d) and vesicles (ve) probably laden with wall material are present.
Chloroplasts (chl) have poor lamellar structure and starch content continues to increase.

plane. Studies with metabolic inhibitors and drugs indicate that the determination of the correct plane of division of the cell plate is an energy-requiring process. In onion this reorientation of the spindle and associated structures usually takes 15–20 minutes.

At telophase, nuclear envelopes form from elements of the endoplasmic reticulum around each nuclear mass and the cell plate continues growth until the guard mother cell is completely divided longitudinally. The young cell plate often has a wrinkled appearance and it is often wider or more bulbous at its junction with the original guard mother cell wall. The events described above are depicted in Fig. 3.6 for the differentiation of a guard mother cell of onion.

The separation of the two daughter guard cells to form a stoma is not completely understood. In many species the wall between

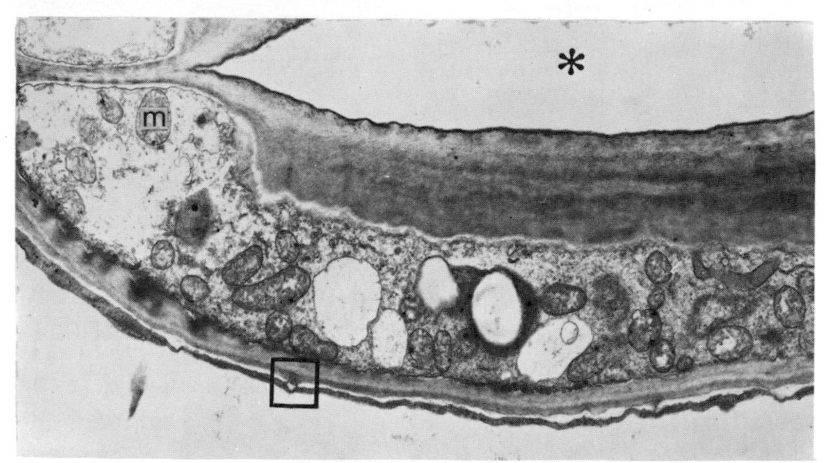

Figure 3.4 (× 8,000) Part of a fully developed stoma of *Phaseolus vulgaris* with open pore (*). Mitochondria (m) are abundant. The box is enlarged in Fig. 3.5.

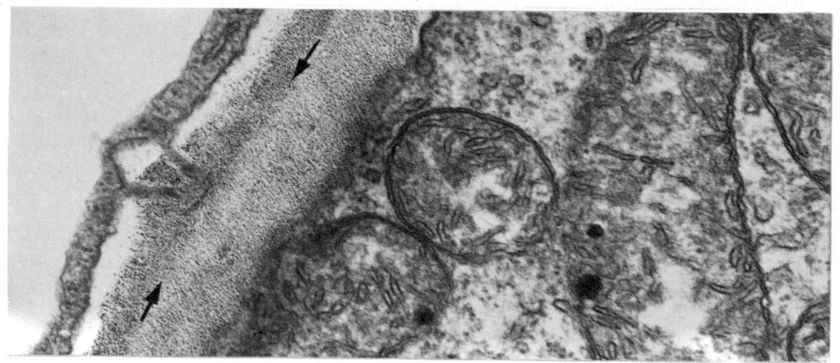

Figure 3.5 The enlarged area (mag. × 50,000), showing incomplete plasmodesmata. The arrows indicate the probable line separating the original primary wall from wall material laid down by the guard cell at a later stage.

daughter guard cells appears to swell and the connection between them weakens. Presumably the breakdown of the pectinaceous middle-lamella occurs due to secretion of pectinases from each of the daughter guard cells. Stevens and Martin (1978) also believe that, at least in *Polypodium*, increased pressure within the guard cells causes the final mechanical separation of the walls which are still attached in the region of what will be the upper and lower lips of the guard cells. Why the common wall between the two guard cells does not split along its entire length is not known. However, a layer of electron-translucent wall material is occasionally observed along this region of wall development and, as in the formation of leaf air spaces, the separation of the wall may be restricted, for unknown reasons, to this region. Guard cell growth may still continue after pore formation along with the development of a substomatal cavity. The substomatal cavity may form by natural extension of existing air spaces and/or by some degeneration of developing mesophyll cells underneath a stoma.

The development of grass-type stomata is considerably different from that observed in elliptical stomata although the initial steps are similar. The guard mother cell divides to form two guard cells which assume a kidney-shaped form and a small pore may form between the cells. A transient swelling of the guard cells then occurs with concomitant elongation followed by a constriction phase which continues until the characteristic dumb-bell-shaped guard cells are formed. The cellulose microfibrils are predominantly axially arranged in the central connecting pieces of the guard cells while the bulbous ends have microfibrils radiating out from the pore (Fig. 3.7).

In Cyperaceae a similar pattern of development occurs to that in the grasses but the microfibrils are orientated radially along the length of the guard cells (Mishkind *et al.* 1981). In *Cyperus esculentus* (yellow nutsedge), however, the alignment of the microfibrils appears to be subject to environmental modification: greenhouse-grown plants have stomata with axial microfibrils in the central connecting pieces of the guard cells while plants growing in the field have radially arranged ones (Mishkind *et al.* 1981). Cortical microtubules direct the alignment of the cellulose microfibrils in the developing cell wall. The microtubules are orientated parallel to the alignment of the cellulose microfibrils but the mechanism by which the microtubules themselves are orientated and the way in which they direct the orientation of the cell wall microfibrils is not clear.

3.2 Classification of stomata

Subsidiary cells (or cells neighbouring the guard cells) may arise from divisions of the guard mother cell or from any other protodermal cell. Based on this ontogenetic relationship of the guard cells and neighbouring cells, three categories have been made. Such classifications require developmental studies because the mature stomatal pattern does not necessarily reveal the ontogenetic relationship of the cells.

Mesogenous: guard cells and neighbouring cells with a common origin.

Perigenous: guard cells and neighbouring cells with no common origin.

Mesoperigenous: at least one neighbouring cell has a common origin with the guard cell. Thus, the stomatal complex is both perigenous and mesogenous.

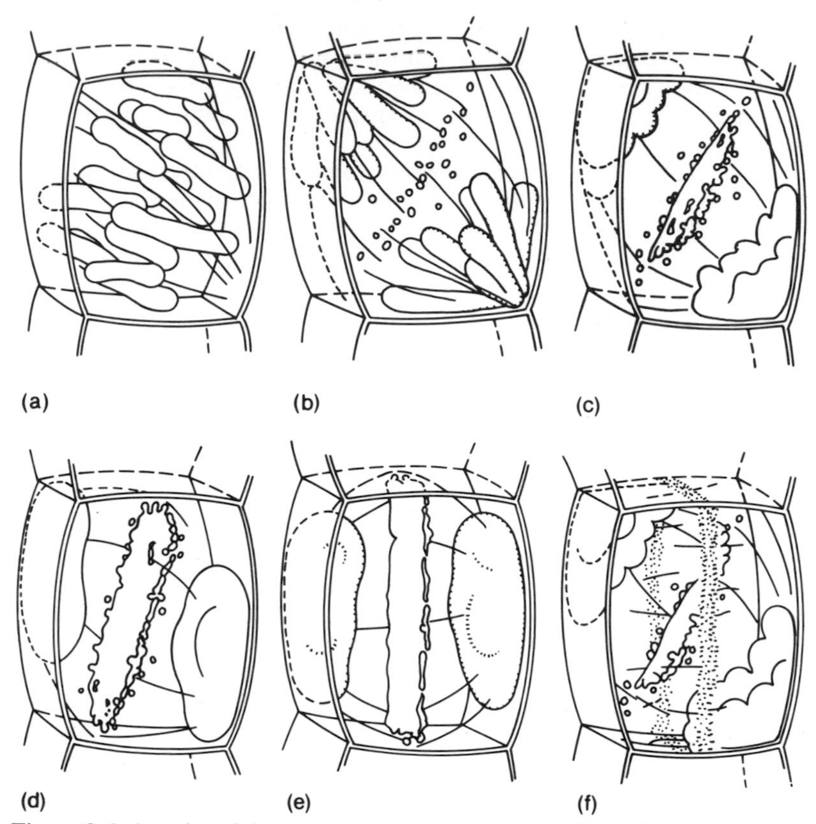

Figure 3.6 A series of drawings depicting mitosis, cytokinensis and reorientation. (a) Metaphase. The chromosomes are aligned between the poles of the obliquely orientated spindle. (b) Late anaphase. The oblique midzone accumulates vesicles. (c) Telophase. The cell plate has begun to form. Reorientation movements have been initiated as daughter nuclei begin to move along the sides of the cell. Note that the plane of the midzone and cell plate in (b) and (c) is canted toward the longitudinal direction and away from the perpendicular to the pole-to-pole axis. Thus the microtubule distance at the edge of interzone at the upper left of the plate is shorter than that at the lower left of the plate. Likewise, the distance at the lower right of the plate is shorter than at the upper right. Force exerted at the upper left and lower right by microtubule growth could contribute to reorientation. (d) Reorientation in progress as plate growth continues. The edges of the plate appear to 'turn in'. (e) Reorientation completed. (f) The same stage as (c), but with a hypothetical position of a morphogenetically cortical site drawn in. Such a site, possibly containing microtubules and indicated by the prior location of the preprophase band, could interact with the late anaphase-telophase spindle and contribute to reorientation (from Palevitz and Hepler 1974).

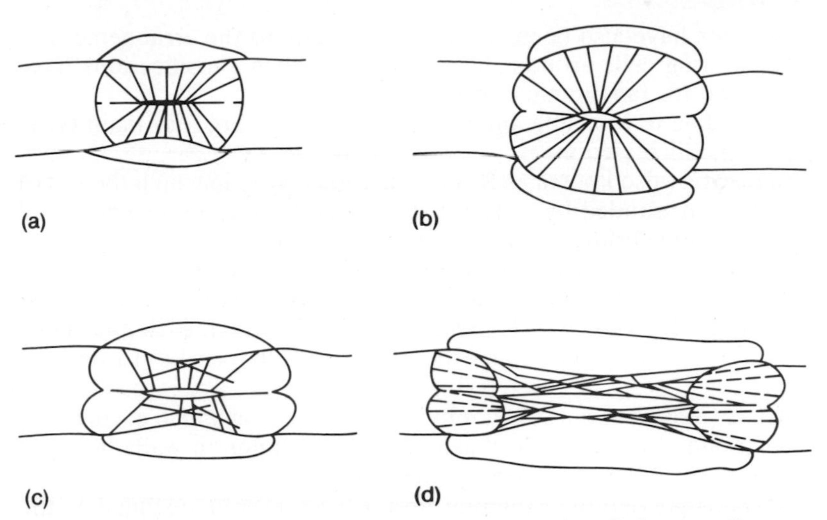

Figure 3.7 Diagrammatic representation of changes in cell shape, wall deposition and microtubule-microfibril orientation in grass guard cells (from Palevitz 1981).

The mesogenous type of development is represented in the *Sedum* genus where the asymmetric divisions of a protodermal cell produce the subsidiary cells before one more division leads to formation of the two guard cells.

Stomatal development in the grasses is of the perigenous type. An unequal division of a protodermal cell produces the guard mother cell as described above. Before the guard mother cell divides to produce the guard cells the contents of the protodermal cells (subsidiary mother cells) on either side of the guard mother cell polarise. The region with the dense cytoplasmic contents and nucleus lies adjacent to the guard mother cell and an asymmetric division of the subsidiary mother cell produces a larger more vacuolated cell destined to become a non-specialised epidermal cell and a smaller cell with dense cytoplasmic contents which develops into a subsidiary cell. As in division of the guard mother cells, a preprophase band of microtubules is also present in subsidiary mother cells in the position of the future cell plate

Plasmodesmata also appear to be concentrated in the common wall between guard mother cells and subsidiary mother cells.

In the development of stomatal complexes in the Commelinaceae the lateral subsidiary cells are always formed first followed by the terminal subsidiary cells before division of the guard mother cell finally occurs.

Stomata have also been classified according to the arrangement of neighbouring cells around the guard cells. Such classifications have been used for taxonomic purposes.

According to Metcalfe and Chalk (1950) there are four main types within the dicotyledons:

Anomocytic (also known as Ranunculaceous type) in which the guard cells are surrounded by cells indistinguishable from other epidermal cells, i.e. no subsidiary cells are present.

Anisocytic (also known as Cruciferous type) in which guard cells are surrounded by three unequally-sized subsidiary cells.

Paracytic (also known as Rubeaceous type) in which each guard cell is accompanied by one or more subsidiary cells parallel to its long axis.

Diacytic (also known as Caryophyllaceous type), in which each stoma is surrounded by two subsidiary cells, their common walls being at right angles to the long axis of the stoma.

In monocotyledons Stebbins and Khush (1961) distinguish the following types:

1. Stoma surrounded by 4–6 subsidiary cells as in the Commelinaceae.
2. Stoma surrounded by 4–6 subsidiary cells of which two are roundish and smaller than the rest and are situated at the ends of the stoma. Examples are found in the Palmae.
3. Stoma which possesses one lateral subsidiary cell per guard cell as in the grasses.
4. Stoma without subsidiary cells as in the Filiaceae. This type is considered more primitive than the rest.

In both the monocotyledons and the dicotyledons, variations on the types described above have been reported and the classifications have been modified according to different authorities (Paliwell 1969; van Cotthem 1970).

3.3 The ultrastructure of guard cells

In the search for clues to broaden our understanding of how stomata function fully developed guard cells have also been the subject of extensive electron microscopic studies. Considerable difficulties have been encountered with such studies due to the small size of guard cells and their location in, often, a thick, tough layer of epidermis; they are also prone to distortion upon preparation for EM studies.

Essentially, guard cells contain dense cytoplasmic contents with relatively small vacuolar volumes. As well as possessing some fascinating and unusual ultrastructural features guard cells possess all of the usual cell organelles (Fig. 3.8). Table 3.1 gives the frequencies of different organelles in guard cells and mesophyll cells of *Allium porrum* (leek) and *Vicia faba* (broad bean).

3.3a Plastids

Until recently it was considered that all functional stomata possessed chloroplasts. However, Nelson and Mayo (1975) reported the absence of chloroplasts in functional guard cells of certain *Paphiopedilum* (lady slipper orchid) species. Also, Jamieson and Willmer (unpublished data) found no chloroplasts in functional guard cells of a variegated chimera of *Pelargonium zonale*.

The number, size and complexity of guard cell chloroplasts varies greatly depending on the species. Two chloroplasts per guard cell are reported in *Anthoceros* while in *Selaginella* three to six chloroplasts per guard cell are found (*see* Allaway and Milthorpe 1976). In most species there are usually about ten to fifteen chloroplasts per guard cell but in ferns chloroplasts are very abundant, with up to 100 reported to pack each guard cell of *Polypodium vulgare*. Figures 3.9 to 3.12 illustrate the wide range of chloroplast frequency and complexity in guard cells of different species, at the light microscopic level.

In most species, guard cell chloroplasts are poorly developed, with little thylakoid structure and scant granal stacking compared to that found in mesophyll cell chloroplasts. In some species of *Allium*, chloroplasts are very small and may not be observed with normal light microscopy although fluorescence microscopy and electron microscopy confirm their presence. Also, in some *Allium* species the granal stacking is extremely sparse in chloroplasts of the guard cells. At the other extreme, chloroplasts in guard cells of ferns are large, with considerable thylakoid complexity and an abundance of starch.

Starch is lacking in the guard cell chloroplasts of a number of species from various families including the Liliaceae, Iridaceae and Amaryllidaceae. In the *Allium* genus starch is lacking in the chloroplasts (Fig. 3.13) but, although there has been some dispute, recent evidence indicates that small molecular weight fructans (polymerisation value of 8 to 9) occur in the epidermis of *Allium cepa*. If fructans occur in the guard cells of the epidermis they could play a similar role to starch, a glucan polymer, by supplying carbon for synthesis of organic anions (see §7.3b).

Guard cell chloroplasts are not destarched by darkness as are chloroplasts of mesophyll cells; instead, starch tends to accumulate in guard cell chloroplasts in the dark. Generally, there is a good negative correlation between guard cell starch content and stomatal aperture (but see §7.3b), starch levels increasing with decreasing stomatal aperture.

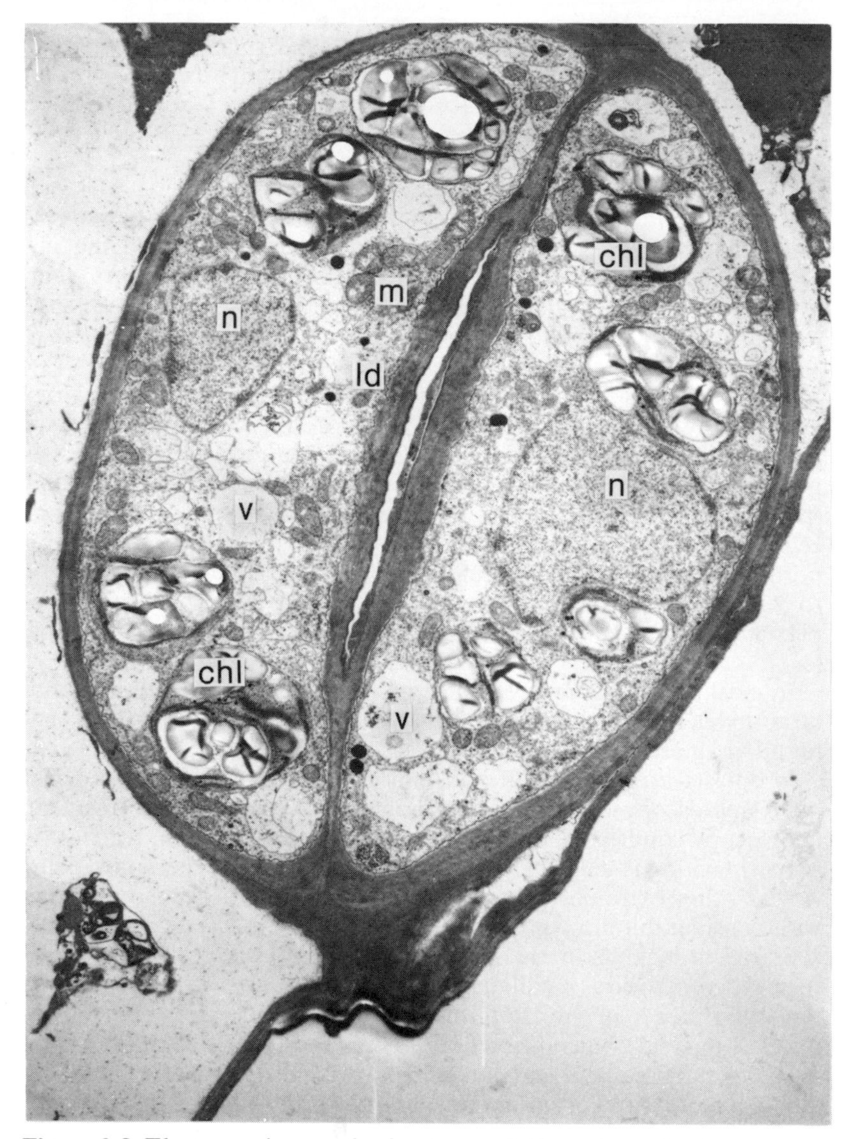

Figure 3.8 Electron micrograph of a stoma of *Phaseolus vulgaris*. (chl) chloroplasts containing starch and poor lamellar structure (n) nucleus, (v) vacuole, (m) mitochondria, (ld) lipid droplets.

Table 3.1 Numbers of organelles per cell profile in thin section

Cell type	Species	Chloroplasts	Mitochondria	Mitochondria: chloroplasts ratio	Microbodies	Spherosomes	No. of cells counted
Mesophyll	*Allium porrum*	17.8 (4–29)	16.8 (9–26)	0.94 (0.41–4.3)	4.2 (2–9)	0.6 (0–3)	5
	Vicia faba	15.2 (11–19)	11.0 (7–16)	0.72 (0.37–1.4)	2.0 (1–4)	0	5
Guard cells	*Allium porrum*	5.3 (2–10)	50.8 (36–68)	9.6 (6.0–20)	0.3 (0–2)	12.2 (7–18)	6
	Vicia faba	6.2 (3–11)	22.6 (7–49)	3.6 (1.9–4.5)	0	8.4 (0–31)	5

Note: Values are means with extreme ranges in parentheses obtained by counting all the profiles of organelles in the whole profile of each cell in a thin section (about 600Å thick). All replicate cells, selected at random, were from different plants (From Allaway and Setterfield, 1972).

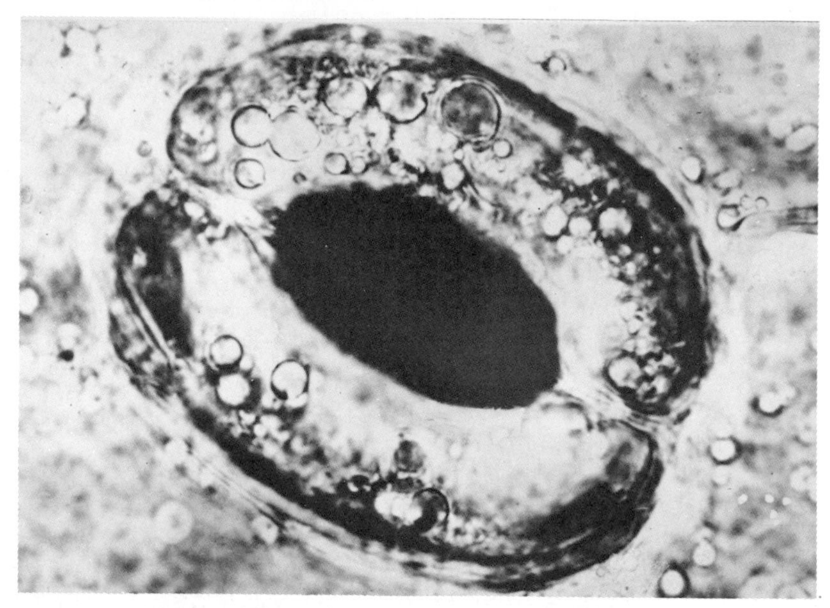

Figure 3.9 *Paphiopedilum aureum hyeanum:* chloroplasts are absent from the guard cells but an abundance of lipid droplets is present (from Rutter and Willmer 1979).

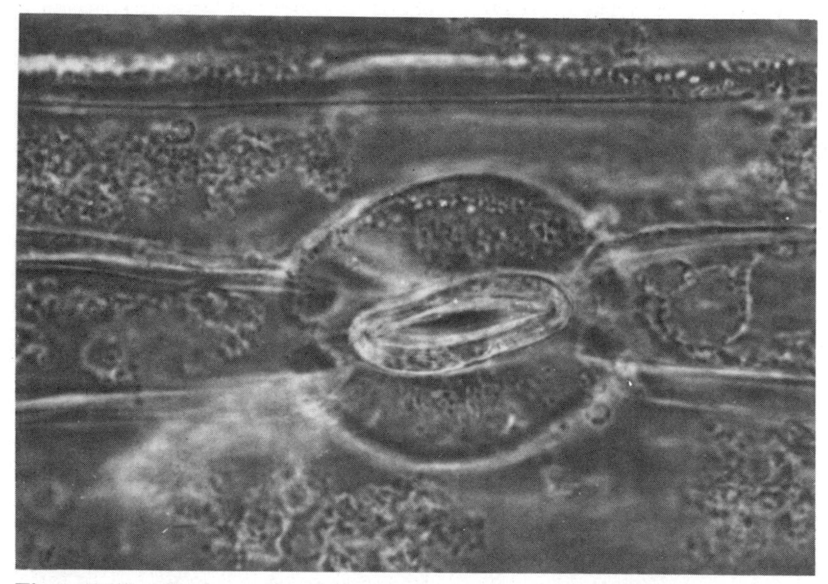

Figure 3.10 *Allium cepa* (onion): chloroplasts are small and poorly developed and are not readily observed under the light microscope with normal illumination.

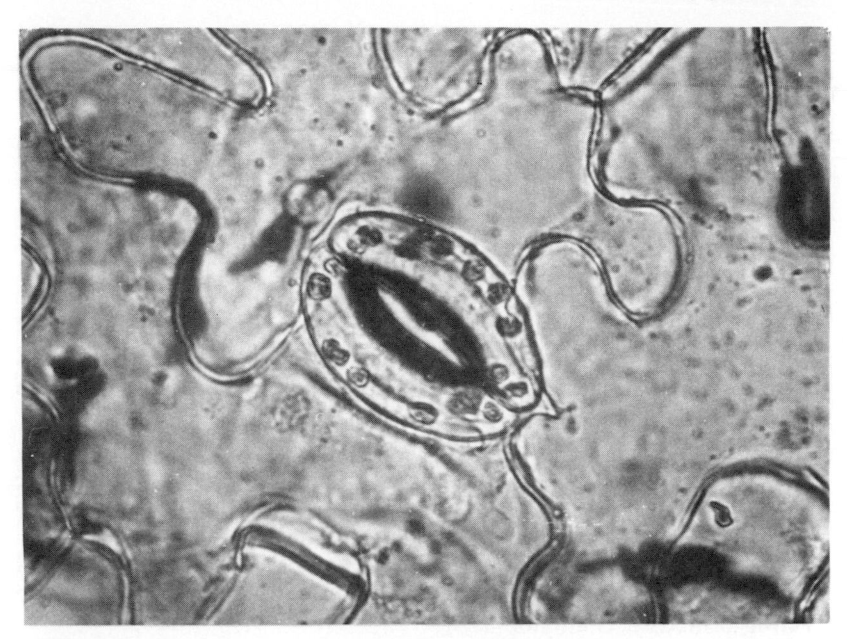

Figure 3.11 *Vicia faba*: the typical situation in which there are relatively few poorly developed chloroplasts in each guard cell.

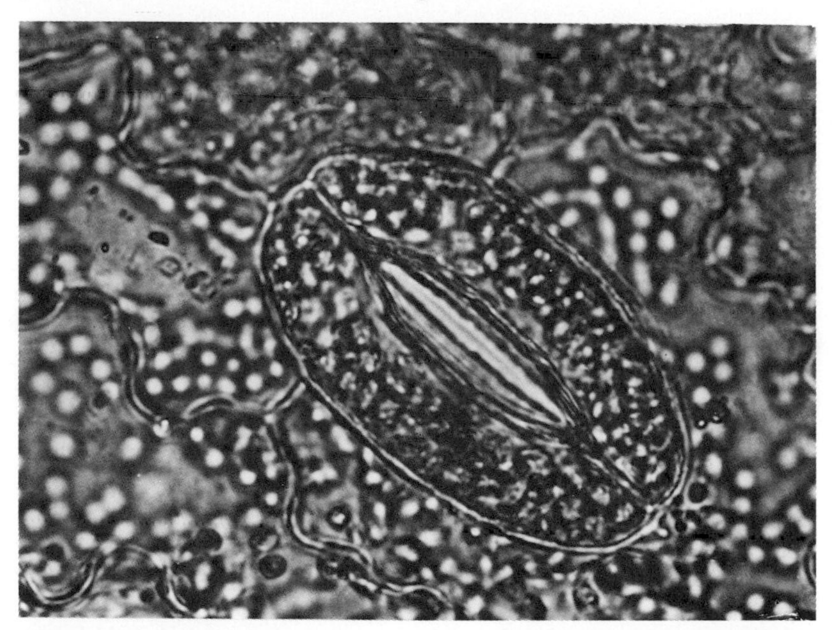

Figure 3.12 *Phyllitis scolopendrium* (Hart's-tongue fern): the guard cells are packed with well developed chloroplasts.

An extensive chloroplast peripheral reticulum (an invagination of the inner of the two bounding chloroplast membranes) has been observed in some guard cells (Fig. 3.13) but this is not a constant feature in all species or within the same species. In C_4 plants, where a peripheral reticulum almost always occurs in the mesophyll and bundle sheath chloroplasts, it has been suggested that one purpose of the reticulum is to increase the transfer of metabolites between cytoplasm and chloroplast. However, since the peripheral reticulum is not a constant feature of guard cell chloroplasts it is obviously not essential for the normal functioning of stomata.

Microtubule-like structures have also been observed in guard cell chloroplasts (Allaway and Setterfield 1972) but their functional significance is not understood. Plastoglobuli are also prevalent in guard cell chloroplasts.

Chloroplasts are lacking in guard cells of *Paphiopedilum insigne*, but another type of plastid is present. These plastids are spindle-shaped with fibrous material running down the long axis. They also contain starch and large, numerous plastoglobuli (Fig. 3.14).

Thus, except for a few species, a common feature of guard cell chloroplasts is the abundance of starch and thus the plastids may be acting primarily as amyloplasts.

3.3b Nuclei

The nucleus of a guard cell is similar in size to that found in many other cell types but since guard cells are relatively small it occupies a greater proportion of the cell volume.

In the kidney-shaped guard cells the nuclei are usually centrally positioned close to the ventral wall. The nucleus may change shape during stomatal movements. For example, the nucleus in guard cells of *Vicia faba* is oval when stomata are closed and rounded when stomata are open. Also, in *Anemia rotundifolia* the nucleus is rounded when stomata are closed and crenated in open stomata.

In dumb-bell-shaped guard cells the nucleus comprises a mass of nuclear material located at each bulbous end connected to each other by a strand of nucleoplasm passing through the central connecting portion of the guard cell. Large pores, 1 μm or more across, are present in the common walls of daughter guard cells of a stoma at each bulbous end (Fig. 3.15). The pores are large enough to allow exchange of organelles between each guard cell pair. Thus, the guard cell pair can be regarded as a single binucleate unit and the pressure relations experienced by the contents of one guard cell will also be experienced by the other, unlike in elliptical stomata.

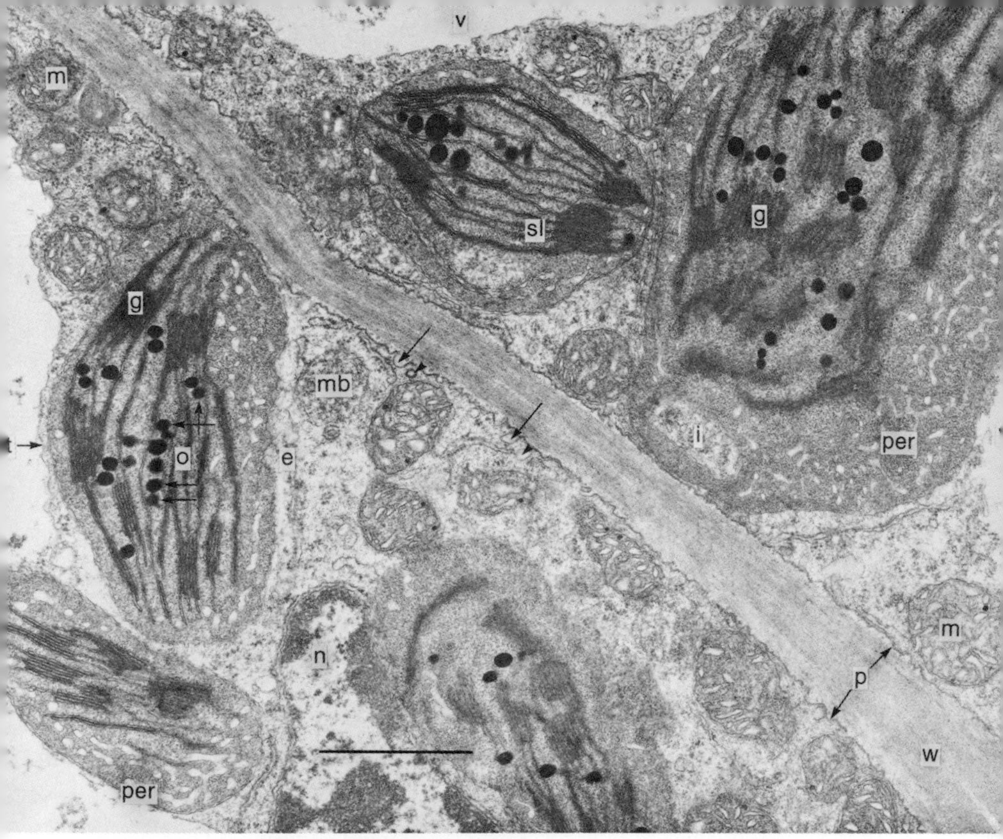

Figure 3.13 (× 20,000; the scale line represents 1 μm) Part of a stoma of *Allium porrum* (leek) showing two guard cells separated by their common ventral walls (w). Note the invaginations of the plasmalemma and vesicles apparently fusing with the cell walls (p and arrowed), numerous mitochondria (m), some endoplasmic reticulum (e), a microbody (mb), chloroplasts [with grana (g), sparse stromal lamellae (sl), numerous osmiophilic globules (o and arrowed), and a well developed peripheral reticulum (per)], vacuole (v) and tonoplast (t). An invagination (i) of the cytoplasm into a chloroplast is also shown (from Allaway and Setterfield 1972).

3.3c Mitochondria

Mitochondria are numerous in guard cells of all species so far examined. This is a feature common to all cells which have high metabolic activity. Their size and shape and appearance of the cristae are not reported to be significantly different from that found in most other cell types.

3.3d Microbodies

Little is known about the microbodies which have been observed in the guard cells of a few species but they are not common and may be

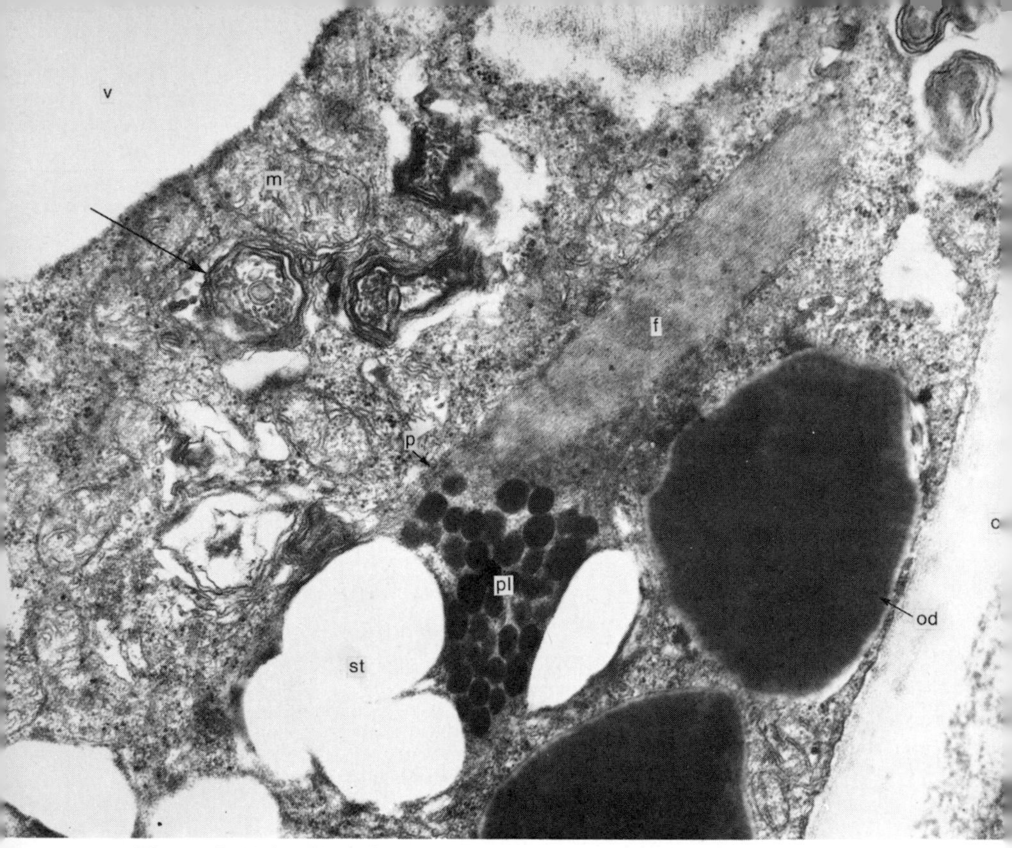

Figure 3.14 (× 75,000) Part of a guard cell of *Paphiopedilum insigne* showing one of the curious plastids (p) which contain starch (st), plastoglubuli (pl) and a fibrous material (f). Also shown are the large oil droplets (od); (v) vacuole; (cw) cell wall; (m) mitochondria. Arrowed are rarely observed membranous structures consisting of concentric whorls of membranes with a tubular or spherical central inclusion (from Rutter and Willmer 1979).

absent in some species. If those microbodies observed in guard cells prove to be peroxisomes, typical of most other green cells, then some glycollate metabolism may occur in the cells since peroxisomes are a central site of glycollate metabolism. However, glycollate oxidase a peroxisomal enzyme activity, is either non-detectable or detectable at only low levels in epidermal tissue and diaminobenzine staining of EM sections of guard cells indicates low catalase (a marker enzyme for peroxisomes) activity. Furthermore, the major source of glycollate in green cells is the Calvin cycle which is not detectable in the guard cells of species investigated so far, including *Vicia faba* (see §7.3b). Thus, it is unlikely that peroxisomes are an important feature in guard cells.

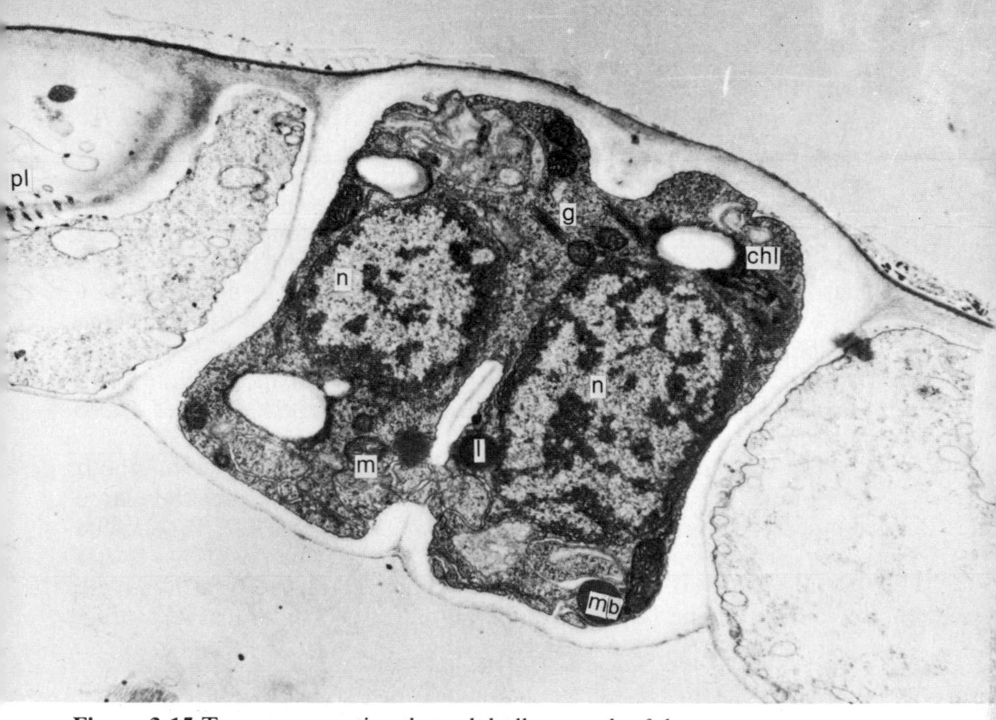

Figure 3.15 Transverse section through bulbous ends of the grass-type stoma of *Leptochloa dubia*. Note the large perforations between the common wall of each cell; (mb) a microbody; (l) lipid droplet; (g) golgi bodies; (mit) mitochondrion; (n) nuclei; (chl) chloroplasts containing starch; (pl) plasmodesmata between epidermal and subsidiary cells which are not observed between guard cells and subsidiary cells (from Pallas and Mollenhauer, 1972).

3.3e Spherosomes and lipid droplets

The function and nature of spherosomes is not clearly understood. They are about one μm in diameter and recent ontogenic studies indicate that they are surrounded by a half-unit membrane rather than a complete unit membrane. Their function may vary depending on their location within a plant, and besides containing lipids they may contain a variety of enzymes including lipases, phosphatases, endopeptidases, RNAase and DNAase. Spherosomes have also been called oleosomes and lipid bodies although the latter name is misleading because lipid droplets exist in the cytoplasm without being surrounded by a membrane or half-unit membrane.

Large numbers of spherosomes have been reported in the guard cells of *Campanula persicifolia* which contained much phosphatase activity when stomata were closed and little activity when they were open (Sorokin and Sorokin 1968). Fujino (1967) also detected ATPase and/or acid phosphatase activity in guard cells of closed stomata but not open ones although the activity was located throughout the cells. Contrary findings were made by Willmer and Mansfield (1970), who detected acid phosphatase and/or ATPase activity in guard cells of both open and closed stomata, though, again, activity was located throughout the cells. Vacuoles are a site of much acid phosphatase activity in other plant cells and it is possible that the phosphatase activity observed by Fujino (1967) and Willmer and Mansfield (1970) was restricted primarily to the vacuoles of guard cells.

Lipid droplets (apparently without a limiting unit or half unit membrane) are regularly observed in guard cells and unusually large numbers occur in species of *Paphiopedilum* (Figs 3.9, 3.14). The purpose of the oil droplets remains obscure but one possibility is that they are involved in the synthesis of cuticle and waxes which may be very thick on the upper lateral wall and lips of guard cells.

3.3f Vacuoles

According to some sources vacuoles are synonymous with lysosomes of animal cells. However, although autophagy has been seen in plant vacuoles it is difficult to accept that guard cell vacuoles are phagocytic in the way that animal lysosomes are. But, like lysosomes, guard cell vacuoles appear to have a high acid phosphatase activity (*see* §3.3e).

The vacuolar volume of guard cells is small compared with most other cell types and has been observed to change its form and size during stomatal movements: Heller *et al.* (1971) observed that the vacuolar volume increased during stomatal opening, while Guyot and Humbert (1970) observed that the vacuole of *Anemia rotundifolia* consisted of many small ones when stomata were closed and fewer, larger ones when stomata opened. Caution is needed, however, in interpreting results obtained from EM studies since it is difficult to fix material without altering vacuolar shapes and sizes. In living guard cells of onion, autofluorescence of the vacuole indicates that it consists of a network of anastomising vesicles.

3.3g Plasmodesmata and ectodesmata

The absence or presence of plasmodesmata connecting mature guard cells with neighbouring cells has been a contentious issue for many years. They are present as unbranched connections, occurring either singly or in small pit fields between guard cells and neighbouring cells as the stomata develop even up to the stages where the stomatal pore is partly formed. However, at maturity the general view is that complete, functional plasmodesmata do not exist between guard cells and neighbouring cells but that they become aborted as wall material is deposited over them during the extensive wall thickening which occurs as guard cells develop (Figs 3.1 to 3.5).

However, one should be aware that there are numerous reports at the light microscope level of plasmodesmata existing between mature guard cells and neighbouring cells although, at the electron microscope level, where identification is less equivocal, there are very few reports of their existence in this location and only one study (Pallas and Mollenhauer 1972) allows critical evaluation. Pallas and Mollenhauer report large numbers (15–20 pores/μm^2) of plasmodesmata in the end walls of guard cells of *Vicia faba* and tobacco; the plasmodesmata were branched and there was no restriction of thickening in the pit fields. This study remains an isolated case and difficult to reconcile with other EM studies. Observations of complete plasmodesmata being observed in mature guard cells at the light microscope level may be explained if one assumes that the histochemical tests used to detect plasmodesmata give positive reactions to aborted connections or to weakened or structurally different areas of wall material.

There are other reasons for expecting plasmodesmata to be absent in mature tissue: during stomatal opening guard cells of some species bulge into neighbouring cells, stretching the dorsal guard cell wall, and also there are reports that guard cells can change their alignment in the epidermis and move slightly above or below the leaf surface during stomatal movements. It is likely that such distortions and movements would damage plasmodesmatal connections if they were present.

It was believed that fingers of cytoplasm extended from the guard cell protoplast through the outer lateral wall to the cuticular region. These structures, known as ectodesmata (Franke 1962), are now considered to be artifacts and may really consist of regions of differently structured or more permeable wall material.

In the absence of plasmodesmata and of features such as transfer cell-like convolutions of the plasmalemma which would increase metabolite transfer it must be assumed that such transport is not essential for stomatal functioning.

3.3h Microtubules, dictyosomes, endoplasmic reticula and other membranous inclusions

All the structures involved in the synthesis, direction and secretion of wall material and the synthesis of proteins have been observed in guard cells particularly in the early stages of the development and differentiation of stomata.

Microtubules are about 24 nm thick with an electron opaque core about 10 nm across. They are built up of protein units called tubulins which polymerise and, perhaps with direct interaction with the plasmalemma, control the alignment of cellulose microfibrils (*see* §3.1).

Rough endoplasmic reticulum (ER) and polysomes are very frequent during the development of guard cells and are also prevalent in mature guard cells. The frequency of these inclusions indicates a high rate of protein synthesis (*see* Fig. 2.10). Smooth ER and dictyosomes are also evident, particularly during the development of guard cells.

Plasmalemmasomes (also known as lomasomes or paramural bodies) are occasionally seen in guard cells. These are infoldings and inrollings of the plasmalemma into the cytoplasm which may enclose vesicular bodies. Their function in guard cells is obscure.

Coated vesicles have been observed in guard cell protoplasts of onion, but their function in this or any other cell type is also obscure.

Other membranous structures including concentric whorls of membrane with tubular or spherical central inclusions (Fig. 3.14) have been occasionally observed in guard cells though, again, their significance is not understood.

Further reading and references

Allaway, W. G. and Milthorpe, F. L. (1976) Structure and Functioning of Stomata, In *Water Deficits and Plant Growth*, vol. IV, pp. 57–102. Academic Press, New York.

Allaway, W. G. and Setterfield, G. (1972). Ultrastructural observations on guard cells of *Vicia faba* and *Allium porrum*, *Can. J. Bot.*, **50**, 1405–13.

Bünning, E. and Biegert, F. (1953). Die Bildung der Spaltöffnungsinitialen bei *Allium cepa*. Z., *Bot.*, **41**, 17–39.

Franke, W. (1962). Ectodesmenstudien. 1. Mitt: Über pilzförmig erscheinende Ektodesmen. Kritische Abhandlung über das Wesen der Ektodesmen, *Planta*, **59**, 222–38.

Fujino, M. (1967). Role of adenosinetriphosphate and adenosinetriphosphatase in stomatal movements, *Sci. Bull. Fac. Educ. Nagasaki Univ.*, No. 18.

Galatis, B. and Mitrakos, K. (1979). On the differential divisions and preprophase microtubule bands involved in the development of stomata of *Vigna sinensis* L., *J. Cell Sci.*, **37**, 11–37.

Guyot, M. and Humbert, C. (1970). Les modifications du vacuome des cellules stomatiques d'*Anemia rotundifolia* Schrad, *C. R. Acad. Sci.*, Ser. D., **270**, 2787–90.

Heller, F. O., Kausch, W. and Trapp, L. (1971). UV-Mikroskopischer Nachweis von Strukturaderungen in Schliesszellen von *Vicia faba*, *Naturwissenschaften*, **58**, 419.

Metcalfe, C. R. and Chalk, L. (1950) Anatomy of the Dicotyledons, vol. 1 Oxford University Press, London and New York.

Mishkind, M., Palevitz, B. A. and Raikhel, N. (1981) Cell wall architecture: normal development and environmental modification of guard cells of the Cyperaceae and related species, *Plant Cell and Environment*. **4**, 319–328.

Nelson, S. P. and Mayo, J. M. (1975). The occurrence of functional non-chlorophyllous guard cells in *Paphiopedilum* spp., *Can. J. Bot.*, **53**, 1–7.

Palevitz, B. A. (1981). The structure and development of stomatal cells, in *Stomatal Physiology* (Eds. P. G. Jarvis and T. A. Mansfield). Cambridge University Press.

Palevitz, B. A. and Hepler, P. K. (1974a) The control of the plane of division during stomatal differentiation in *Allium*. 1. Spindle reorientation, *Chromosoma*, **46**, 297–326.

Palevitz, B. A. and Hepler, P. K. (1974b) The control of the plane of division during stomatal differentiation in *Allium*. 11. Drug studies, *Chromosoma*, **46**, 327–41.

Paliwell, G. S. (1969). Stomatal ontogeny and phylogeny. 1. Monocotyledons, *Acta Bot. Neerl.*, **18**, 654–68.

Pallas, J. E. and Mollenhauer, H. H. (1972) Physiological implications of *Vicia faba* and *Nicotiana tabacum* guard cell ultrastructure, *Amer. J. Bot.*, **59**, 504–14.

Pickett-Heaps, J. D. (1969). Preprophase microtubules and stomatal differentiation in *Commelina cyanea*, *Austr. J. Bio. Sci.*, **22**, 375–91.

Rutter, J. M. and Willmer, C. M. (1979) A light and electron microscopy study of the epidermis of *Paphiopedilum* spp. with emphasis on stomatal ultrastructure, *Plant Cell and Environment*, **2**, 211–19.

Sorokin, H. P. and Sorokin, S. (1968) Fluctuations in the acid phosphatase activity of spherosomes in guard cells of *Campanula persicifolia*, *J. Histochem. Cytochem.*, **16**, 741–802.

Stebbins, G. L. and Khush, G. S. (1961). Variations in the organisation of the stomatal complex in the leaf epidermis of monocotyledons and its bearing on their phylogeny, *Amer. J. Bot.*, **48**, 51–9.

Stebbins, G. L. and Shah, S. S. (1960). Developmental studies of cell differentiation in the epidermis of monocotyledons. 11. Cytological features of stomatal development in the Gramineae, *Develop. Biol.*, **2**, 477–500.

Stevens, R. A. and Martin, E. S. (1978) Structural and functional aspects of stomata. 1. Developmental studies in *Polypodium vulgare*, *Planta*, **142**, 307–16.

Van Cotthem, W. R. J. (1970). A classification of stomatal types, *Bot. J. Linn. Soc.*, **63**, 235–46.

Willmer, C. M. and Mansfield, T. A. (1970). Further observations of cation-stimulated stomatal opening in isolated epidermis, *New Phytol.*, **69**, 639–45.

Willmer, C. M. and Sexton, R. (1979) Stomata and plasmodesmata, *Protoplasma*, **100**, 113–24.

4 The mechanics of stomatal movements

Although there is not complete agreement on the subject, much is now known about the mechanics and the physico-chemico processes involved in stomatal movements. This chapter deals with such topics and particularly emphasises the water relations of guard cells and epidermal tissue and the different modes of deformation of guard cells during stomatal movements.

4.1 Water relations of guard cells

The motive force for the opening of stomata is the increased turgor pressure of the guard cells (von Mohl 1856; Heath 1938) (*see* Fig. 4.1). Recently, Edwards and Meidner (1979) have measured guard cell turgor pressures directly with inserted pressure probes. The values they obtained for *Tradescantia virginiana* were 0.1 MPa* (1 bar) for guard cells of nearly closed stomata and about 0.7 MPa (7 bar) for guard cells of stomata open to 26 μm. These values are surprisingly low but they may be a feature peculiar to *T. virginiana* since much higher turgor pressures for guard cells of *Commelina communis* have been obtained (Meidner, private communication).

The increased turgor results from the decreased osmotic potential of the guard cells. (Guard cell osmotic potential is used loosely and refers more accurately to the osmotic potential of the vacuolar solutes of the guard cells.) In exceptional cases it has been reported that the osmotic potential of the guard cells remains fairly constant while the osmotic potential of epidermal cells increases (Meidner and Mansfield 1968, p. 21).

Table 4.1 gives a list of guard cell osmotic potentials for open and closed stomata of a variety of species. Although the values will vary

* 1 MPa = 10 bar = 9.87 atomspheres = 100 J kg^{-1}
1 Pa = 1 Nm^{-2} = 1 kg m^{-1} s^{-2}

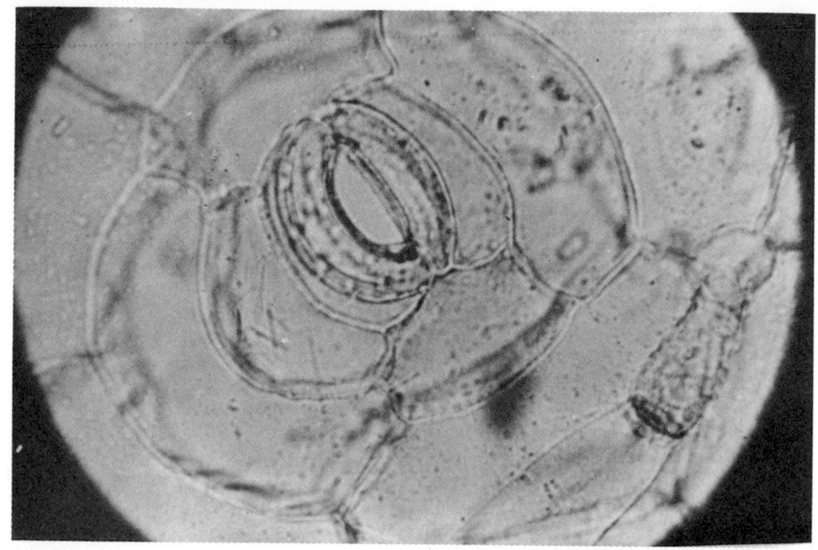

Figure 4.1 A stoma of *Commelina communis* with one punctured guard cell which has collapsed. The intact guard cell retains its shape and internal pressure and the pore adjacent to this cell remains open. This indicates that the polar ends of the guard cels do not necessarily act as firm platforms for each other as their turgor pressures increase to open the stoma.

according to the extent of stomatal opening and according to species many of the published values appear too high (i.e. the negative values are too small) to account for measured concentrations of osmotica in guard cells assuming that the solutes remain dissociated. The reasons for these high values stem from the wrong usage of the incipient plasmolysis method for determining osmotic potentials of guard cells. Errors in the use of the technique can occur due to cell volume shrinkage during plasmolysis which is not accounted for, exosmosis phenomena occurring in the unstirred layers of the plasmolyticum surrounding the cells, entry of plasmolytica into the cells, and solute leakage from the cells during plasmolysis (Fischer 1973; Willmer and Beattie 1978).

Cryoscopic methods have been used to determine osmotic potentials of guard cells: Bearce and Kohl (1970) found much lower values for guard cells of *Chrysanthemum* and *Geranium* than had hitherto been generally obtained, although values varied greatly at each stomatal aperture (e.g. when stomata of *Chrysanthemum* were closed guard cell osmotic potentials ranged from −1.2 MPa to −4.6 MPa) and the relationship between guard cell osmotic potential or the pressure differences between guard cells and epidermal cells and stomatal aperture was not clear. Slade and Willmer (unpublished

Table. 4.1 Osmotic potentials of guard cells and neighbouring epidermal/subsidiary cells when stomata are open or closed

	Guard cells (MPa)		Epidermal/ subsidiary cells (MPa)		
Stomatal aperture*	**Open**	**Closed**	**Open**	**Closed**	**Reference**
Species					
Veronica beccabunga	−2.5	−2.0	−1.2	−1.2	
Allium cepa	−1.8	−1.5	−0.5	−0.5	
Rumex patientia	−1.9	−1.3	−1.4	−1.3	Meidner and Mansfield (1968) (rows *Veronica beccabunga* to *Vicia faba*)
Cyclamen	−2.2	−1.3	−1.1	−1.1	
Ranunculus bulbosus	−1.4	−1.5	−0.9	−1.9	
Vicia faba	−1.0	−0.9	−0.6	−0.6	
Vicia faba	−3.7(12)	−1.7(2)	−0.6(12)	−0.6(2)	Raschke (1979)
Vicia faba	−4.5	—	—	—	Raschke (1975)
Vicia faba	−1.7(14)	−0.4(4)	—	—	Allaway and Hsiao (1973)
Tradescantia virginiana	−1.0(14)	−0.6	−0.4(14)	−0.3	Meidner and Edwards (1975)
Commelina communis	−2.8	−0.8	−0.2	−0.3	Meidner and Edwards (1975)

* Numbers in parentheses are the stomatal apertures in μm; where parentheses are lacking values are not available but are assumed to be 0 for closed stomata.
— No figures available.

data) have also used cryoscopic methods to measure osmotic potentials of guard cells of *Tradescantia virginiana* and found much lower values than had previously been obtained by the incipient plasmolysis method. However, further work is needed to fully evaluate the accuracy of the cryoscopic method particularly since there is some evidence that colloids do not behave as dissolved solutes do in terms of depression of freezing points.

Since the guard cell tonoplast is differentially permeable and has little mechanical strength it is considered that the cytoplasm must adjust osmotically to match the osmotic potential of the vacuole as stomata open and close. Some possible ways to decrease the osmotic potential of the guard cell cytoplasm are: (1) to dehydrate the cytoplasm resulting in decrease of its volume; (2) for the cytoplasm to retain some of the K^+ and Cl^- as the ions move toward the vacuole; (3) for the cytoplasm to accumulate from outside, or synthesise from within, considerable amounts of organic solutes such as betaines, amino acids or sugars. There could also be some combination of these three mechanisms operating together. Evidence to date, however, suggests that changing levels of a sucrose pool could be involved as a cytoplasmic osmoticum.

It is evident that the turgor pressures of guard cells do not balance the values generally obtained for the osmotic potentials of guard cells (Fig. 4.2). This is an apparent anomaly since, conventionally, the total water potential of cells or tissues at full turgor is essentially equal to the sum of the solute potential, ψ_s, the pressure potential (hydrostatis turgor pressure), ψ_p, and the matric potential, ψ_m, which is commonly included with the solute potential and called osmotic potential, ψ_π.

Thus, $\psi = \psi p + \psi_\pi$

ψ_π can only be negative, while ψ_p can only be zero or positive (some authorities consider that under special circumstances negative ψ_p can occur). Thus, when guard cells are fully turgid, ψ_π should be close to ψ_p. As already indicated, this is not the case. In partial explanation of this Edwards and Meidner (1979) consider that, at least for open stomata, guard cells may never reach full turgor because water evaporates from the guard cells rapidly enough to create low matric potentials in the walls and these compete strongly against the low solute potentials of the guard cell protoplast for water.

4.2 Reflection coefficients and osmotic effects of ions

The reflection coefficient (σ) is a measure of the nonpenetration of a solute through a membrane that is freely permeable to water. If a membrane is completely permeable to a solute, the reflection coefficient of that solute is zero while, if the membrane is completely impermeable to a solute, the reflection coefficient is one. Thus values between one and zero are obtained for different solutes.

Many ions, particularly K^+, readily enter guard cells, indicating that they have a low reflection coefficient with values near zero. If these ions can readily enter guard cells they should also be able to leave readily. Thus, the contribution of K^+ to the decreased osmotic potential of guard cells has been questioned (Gradmann 1977). However, as Hastings and Gutnecht (1978) and Raschke (1979) have pointed out the reflection coefficient of an ion is determined by the least permeable of the ions within a solution due to tight electrical attraction of cations and anions. In guard cells K^+, some Cl^-, and organic acid anions are the major ions to accumulate in vacuoles of guard cells as stomata open. These ions, plus fixed charges in guard cells, will all determine the mobility of a specific ion within the cells. Thus, if the permeability of the tonoplast is low for Cl^- or organic acid anion (i.e. these ions have high reflection coefficients) then K^+ will remain in the guard cell vacuole and contribute to the guard cell osmotic potential.

4.3 Phases of stomatal opening

Stålfelt (1927) proposed two phases of stomatal opening, namely, 'Spannungsphase' (tension phase) and 'motorische Phase' (motor phase). Another phase may be termed the 'aperture maintenance phase' in which the stomatal aperture reaches a certain level and then remains constant (Rogers *et al.* 1979). 'Spannungsphase' is a period preceding the motor phase in which guard cells ready themselves for stomatal opening; guard cells inflate due to turgor increases but the pore does not open, or opens to only a tiny extent. The motor phase is that phase in which there is an approximately linear relationship between pore opening and guard cell turgor (Fig. 4.2).

A linear relationship between guard cell volume and guard cell osmotic potential also exists in a number of species (Fig. 4.3). A mathematical model, using information from *Vicia faba* stomata, predicted a nearly linear relationship although, additionally, a pronounced inflection of the line occurred at about 1 μm aperture presumably as the transition from Spannungsphase to the motor phase occurred (Fig. 4.3).

4.4 The mechanical advantage of neighbouring cells over guard cells

A common misconception about the pressure relations of stomatal movements is that the pore aperture is a direct function of the

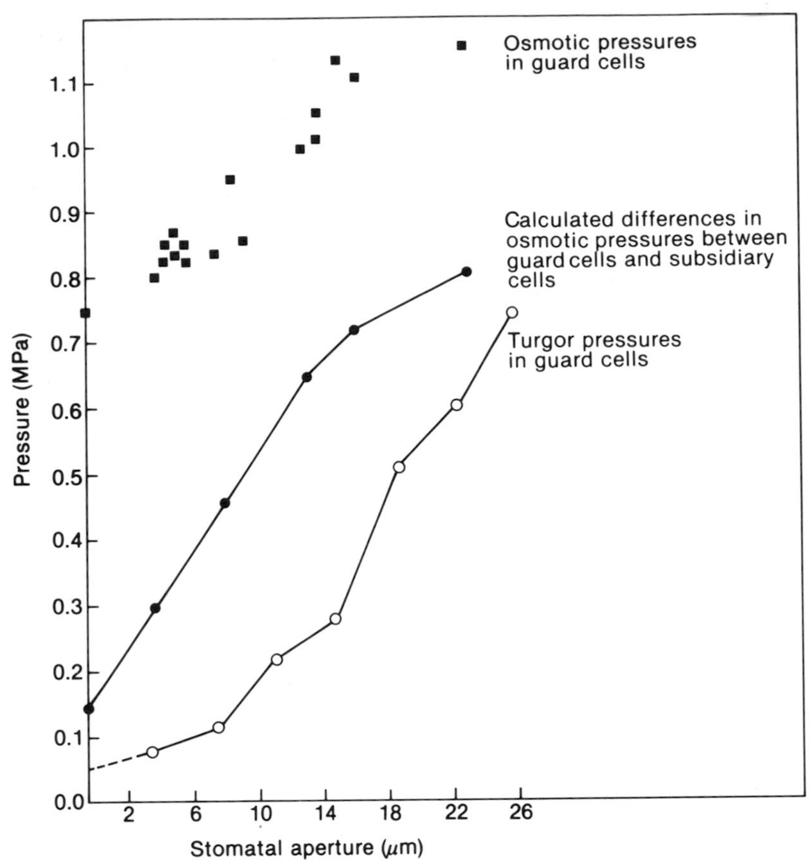

Figure 4.2 The relationship between guard cell osmotic pressure (■), the difference in osmotic pressure between guard cells an subsidiary cells (●), and guard cell turgor pressure (o) and stomatal aperture in *Tradescantia virginiana*. (Adapted from Edwards and Meidner 1979). The osmotic pressure values were determined at incipient plasmolysis and may therefore be underestimates due to cell shrinkage.

pressure differences between guard cells and neighbouring cells. However, neighbouring epidermal or subsidiary cells have a 'mechanical advantage' over the guard cells (von Mohl 1856; De Michele and Sharpe 1973; Edwards *et al*. 1976) and pressures in guard cells must build up to a certain level above pressures in neighbouring cells before the stomatal pore begins to open. The period of build-up of guard cell pressure corresponds to the Spannungsphase of stomatal opening. This mechanical interaction between guard cells and neighbouring cells has also been referred to as the antagonism ratio (Cooke *et al*. 1976).

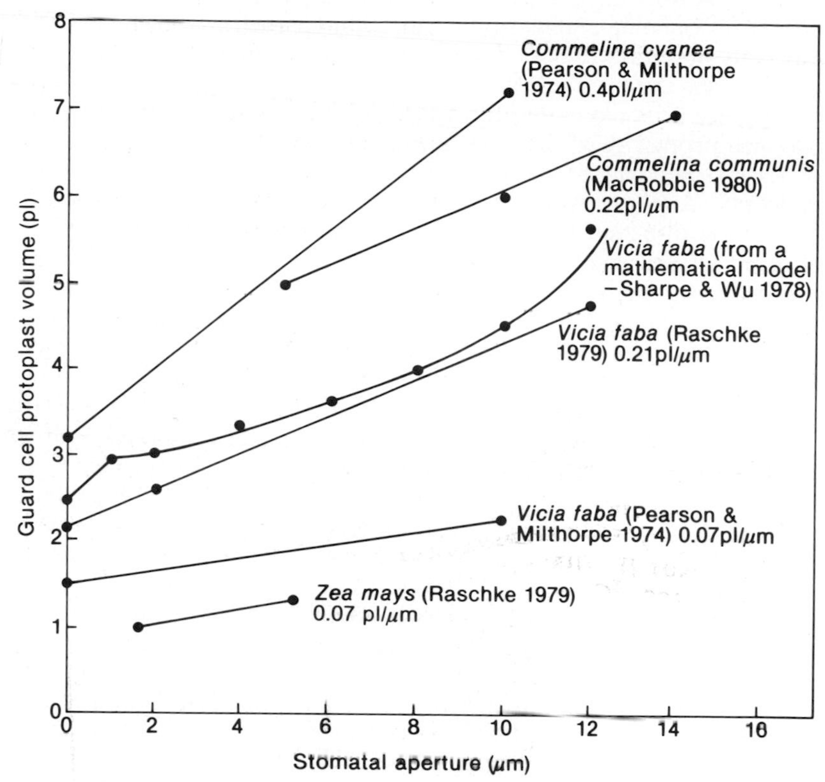

Figure 4.3 The relationship between guard cell protoplast volume and stomatal aperture for different species. Associated with each line is the name of the species, the source of the information and the increase in volume per μm increase in stomatal aperture.

4.5 Deformation of guard cells during stomatal movements

Numerous models describing the mechanics of stomatal opening and closing have been formulated and assembled by various workers (De Michele and Sharpe 1973; Aylor *et al.* 1973; Cooke *et al.* 1976). However, there are considerable difficulties in comparing non-living, physical structures, such as steel beams, with the more dynamic biological materials, such as cell walls, which surround living cells. This is particularly the case for guard cells in which not only may the walls change their thickness during stomatal movements but the wall properties may also change due to binding and release of different ion species as opening and closing occurs. Under such circumstances the cell wall material will not obey Hooke's law, a fact not allowed for in some of the earlier models.

Another problem is that the guard cell wall is anisotropic, i.e. the physical properties of the wall (particularly of stretching and elasticity) are dependent on the direction of measurement, as opposed to isotropic structures which have uniform physical properties regardless of the direction in which they are measured. This anisotropy is a result of the micellar structure of the cell walls and the non-uniformity of the wall material – different areas of the wall may have different amounts of pectin, cellulose and hemicelloses, for example.

Additionally, stomatal movements are controlled, to some extent, not only by the turgor and rigidity of the epidermal cells but also by the underlying mesophyll, particularly if the mesophyll is closely attached to the epidermis.

The extent and location of extra wall thickening around guard cells can vary greatly from species to species. It is possible, therefore, that there are also considerable variations in the mechanics of the deformation of guard cells during stomatal movements. However, it is extremely difficult to make the necessary measurements of guard cell dimensions at different stomatal apertures from fresh or freeze sectioned material. Furthermore, reliable information on guard cell dimensions cannot be obtained from EM studies since the fixation and resin embedding procedures result in contraction and some deformation of guard cells. It is considered, therefore, that little benefit can be gained from enumerating and describing all the different guard cell deformations which have been suggested to occur although the reader should be aware of these (see Allaway and Milthorpe 1976). Instead, three main types will be described.

1. The simplest type is reported to be in the mosses such as *Mnium cuspidatum* (Haberlandt 1896) and possibly some ferns in which the stomatal pore forms as a result of the guard cells becoming narrower and deeper; upon closure the ventral walls become more rounded and the guard cells wider and less deep. Only the ventral walls of the guard cells increase in length as the stomata open. Figure 4.4 illustrates these deformities.

2. Guard cell deformation of other elliptical-shaped stomata have certain features in common which allow them to be discussed together. Although it has occasionally been questioned the length of a stoma remains virtually constant during opening and closing movements. The guard cells, however, increase in length upon stomatal opening: both the ventral and dorsal walls increase by about the same absolute amount but, since the ventral wall is shorter than the dorsal wall, the former will extend relatively more than the latter (Raschke 1975; Meidner and Willmer 1975) (Fig. 4.5). Thus, the thick ventral walls *are* stretchable. The radial orientation of the

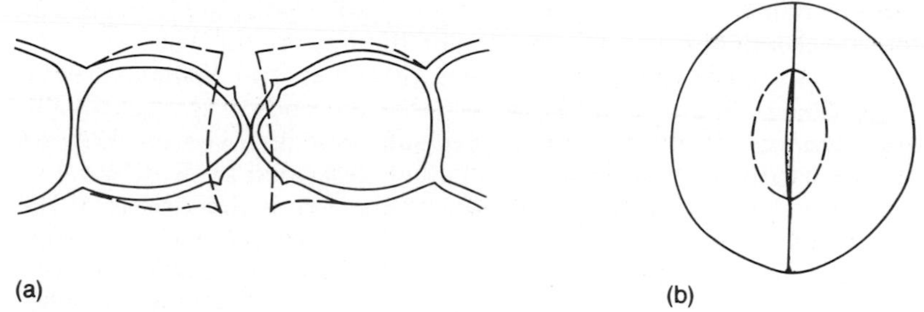

Figure 4.4 Deformation of guard cells of *Mnium cuspidatum* (A) side view, (B) plan view (-----stoma open;——stoma closed).

cellulose microfibrils will allow expansion of the guard cells in a lengthwise direction. (It is not known, however, if this is a common feature of all kidney-shaped guard cells.) Such a radial arrangement of the microfibrils will allow the cross sectional shape to change but the cross sectional area of the outer dimensions of the guard cells will remain almost constant. Using rubber balloon models Aylor *et al.* (1973) demonstrated the importance of radial micellation of the guard cell walls.

According to Raschke (1975, 1979) opening is effected in the following manner: the ends of the guard cells swell and push each other apart and, since the length of the stomatal apparatus remains virtually constant, guard cell expansion occurs outwards into the surrounding epidermal or subsidiary cells. However, when one guard

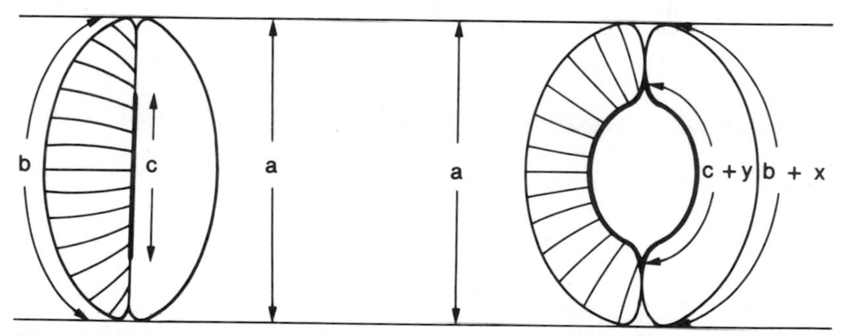

Figure 4.5 Open and closed stomata showing that the overall length, (a), does not change during stomatal movements but the length of the dorsal, (b), and ventral, (c), walls of the guard cells increase by distances x and y, respectively, upon stomatal opening. The radial direction of the cellulose microfibrils are also diagrammatically represented in one guard cell of each stoma.

cell of a pair is punctured the intact guard cell is still able to expand in the usual manner and this tends not to support the view that the guard cells push against each other (Meidner and Willmer 1975). Nevertheless, it may still be possible that the ends of the guard cells are anchored firmly enough to give sufficient leverage to allow an intact guard cell to swell in the manner described by Raschke.

Although the cross sectional area of the outer dimensions of the guard cell may not change (in those species having the radial arranged microfibrils) during stomatal movements, Raschke (1979)

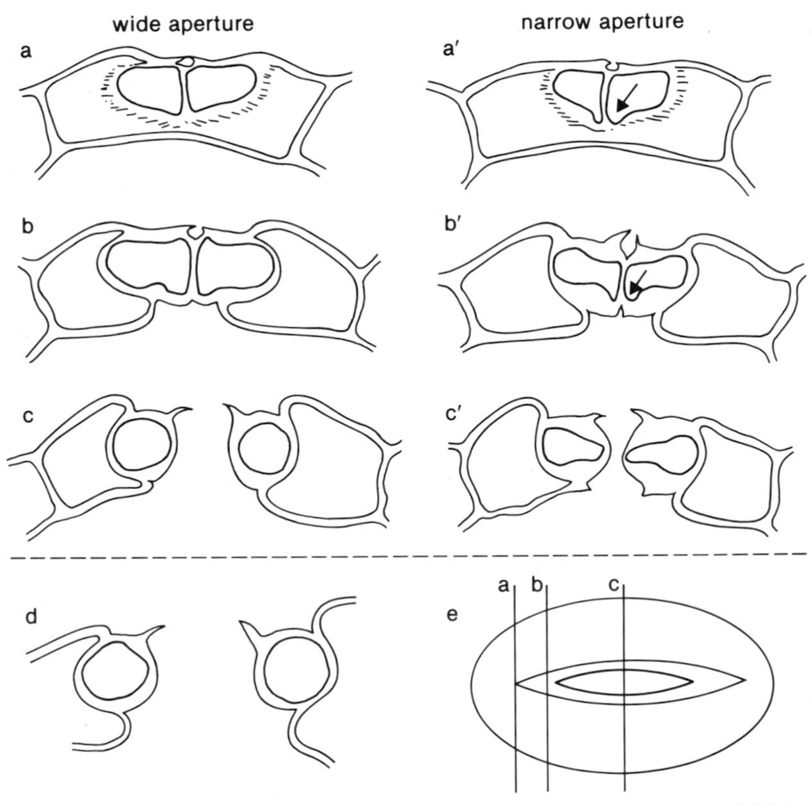

Figure 4.6 Tracings of optical cross sections through an intact stoma of *Vicia faba* in a wide (10.5 μm)) and narrow (1.5 μm) state of aperture. The stoma was first photographed in the open state and then plasmolyzed with a sucrose solution. Approximate position of the focal planes is shown in (e). (d) is a median section through another stoma which had opened to an aperture of 21.2 μm. Arrows point to the fold that forms in the polar parts of the guard cell when the stoma closes (from Raschke 1979).

considers that the cross sectional area of the protoplast does change (Fig. 4.6). According to Raschke, when stomata are closed or nearly closed the cross sectional shape of the middle portion of the protoplast is triangular in shape but upon stomatal opening becomes more spherical. Changes also occur in the cross sectional shape of the end regions of the guard cells during stomatal movements: in the bottom of the ventral wall the protoplast exists as a ridge when stomata are closed which is stretched out as turgor increases and stomata open. Such changes in protoplast shape and volume would result in considerable decreases in the thickness of the wall of the guard cell.

3. The wall thickenings of grass-type guard cells are very different from those found in kidney-shaped guard cells. The thin-walled ends of the dumb-bell-shaped guard cells swell as the stoma opens and are able to push the middle portions apart (Fig. 4.7). The middle portions of the guard cells are heavily thickened on the upper and lower surfaces of the lateral walls and these are relatively inflexible. As a result a pore is formed with parallel sides. According to Raschke (1979) the bulbous ends of the guard cells act like bellows pushing the mid-portions apart and, thus, resemble the system that he envisages in elliptical stomata.

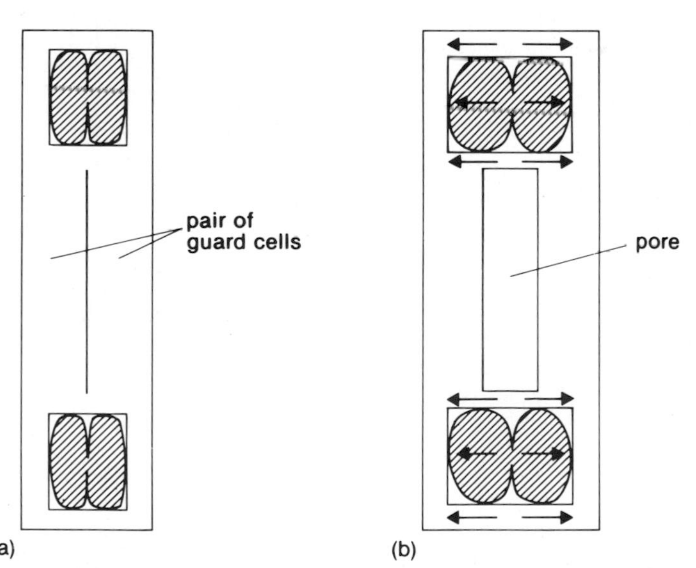

Figure 4.7 Diagrammatic representation of possible deformations of guard cells in a grass-type stoma, (A) closed, (B) open. The protoplasts in the bulbous ends of the guard cells, which are connected to each other via pores, swel up and push the guard cells apart in the direction indicated by the arrows.

Further reading and references

Allaway, W. G. and Hsiao, T. C. (1973). Preparation of rolled epidermis of *Vicia faba* L. so that stomata are the only viable cells: analysis of guard cell potassium by flame photometry, *Aust. J. Biol. Sci.*, **26**, 309–18.

Allaway, W. G. and Milthorpe, F. L. (1976). Structure and functioning of stomata, in *Water Deficits and Plant Growth*, vol. IV. Academic Press Inc. New York.

Aylor, D. E., Parlange, J. Y. and Krikorian, A. D. (1973). Stomatal mechanics, *Amer. J. Bot.*, **60**, 163–71.

Bearce, B. C. and Kohl, H. C. Jr. (1970). Measuring osmotic pressure of sap within live cells by means of a visual melting point apparatus, *Plant Physiol.*, **46**, 515–19.

Cooke, J. R., DeBaerdemaeker, J. G., Rand, R. H. and Mang, H. A. (1976). A finite element shell analysis of guard cell deformation, *Trans. ASAE*, 1107–21.

DeMichele, D. W. and Sharpe, P. I. H. (1973). An analysis of the mechanics of guard cell motion, *J. Theor. Biol.*, **41**, 77–96.

Edwards, M. and Meidner, H. (1979). Direct measurements of turgor pressure potentials, *J. Exp. Bot.*, **30**, 829–37.

Edwards, M., Meidner, H. and Sheriff, D. W. (1976). Direct measurements of turgor pressure potentials of guard cells. II. The mechanical advantage of subsidiary cells, the Spannungsphase and the optimum leaf water deficit, *J. Exp. Bot.*, **27**, 163–71.

Fischer, R. A. (1973). The relationship of stomatal aperture and guard cell turgor pressure in *Vicia faba*, *J. Exp. Bot.*, **24**, 387–99.

Gradmann, D. (1977). Potassium and turgor pressure in plants, *J. Theor. Biol.*, **65**, 597–9.

Haberlandt, G. (1896). *Physiologische Pflanzenanatomie.* 2nd edn, Englemann, Leipzig.

Hastings, P. F. and Gutnecht, J. (1978). Potassium and turgor pressure in plants, *J. Theor. Biol.*, **73**, 363–6.

Heath, O. V. S. (1938). An experimental investigation of the mechanism of stomatal movement with some preliminary observations upon the response of the guard cells to 'shock', *New Physiol.*, **37**, 385–95.

Macrobbie, E. A. C. (1980). Osmotic measurements on stomatal cells of *Commelina communis* L., *J. Membrane Biol.*, **53**, 189–98.

Meidner, H. and Edwards, M. (1975). Direct measurements of turgor pressure potentials of guard cells. 1, *J. Exp. Bot.*, **26**, 319–30.

Meidner, H. and Mansfield, T. A. (1968). *Physiology of Stomata.* McGraw-Hill, New York.

Meidner, H. and Willmer, C. M. (1975). Mechanics and metabolism of guard cells, *Curr. Adv. Plant Sci.*, **17**, 1–15.

Mohl, H., von (1856). Welche Ursachen bewirken Erweiterung und Verengung der Spaltöffnungen? Bot. Z., **14**, 697–704, 713–21.

Outlaw, W. H. Jr and Lowry, O. H. (1977). Organic acid and potassium accumulation in guard cells during stomatal opening. *Natl. Acad. Sci., USA*, **74**, 4434–8.

Pearson, C. J. and Milthorpe, F. L. (1974). Structure, carbon dioxide fixation and metabolism of stomata, *Aust. J. Plant Physiol.*, **1**, 221–36.

Raschke, K. (1975). Stomatal action, *Ann. Rev. Plant Physiol.*, **26**, 309–40.
Raschke, K. (1979). Movements of stomata, in Hampt, W. and Feinleib, M. E. (eds), *Encyclopedia of Plant Physiology*. vol. 7, *Physiology of Movements*. Springer-Verlag, Berlin.
Rogers, C. A., Powell, R. D. and Sharpe, P. J. M. (1979). The relationship of temperature of stomatal aperture and potassium accumulation in guard cells, *Plant Physiol.*, **63**, 388–91.
Sharpe, P. J. H. and Wu, H. I. (1978). Stomatal mechanics: volume changes during opening, *Plant Cell and Environment*, **1**, 259–68.
Stålfelt, M. G. (1927). Die photischen Reaktionen in Spaltöffnungsmechanismus, *Flora*, **121**, 236–72.
Willmer, C. M. and Beattie, L. N. (1978). Cellular osmotic phenomena during stomatal movements of *Commelina communis*. I. Limitations of the incipient plasmolysis technique for determining osmotic pressures, *Protoplasma* **95**, 321–332.

5 Stomatal responses to environmental factors

5.1 Interaction of environmental factors which control stomata

Stomatal movements are directly affected by light quantity and quality, water availability (plant–soil water status), water vapour pressure deficit (atmospheric humidity), CO_2 concentration and temperature; some of these factors and wind movement may also affect stomata in an indirect manner. All of these environmental factors tend to interact, making studies of the effects of one factor on stomatal movements difficult. For example, increasing leaf irradiance will tend to increase the leaf temperature which, in turn, will lower the leaf water potential; the increased temperature may also change the intercellular CO_2 concentrations by changing rates of photosynthesis, respiration and photorespiration. Thus, all factors may be in play at one time so that stomatal aperture is a resultant of all these factors (*see* Fig. 5.1 which shows the interaction of CO_2 concentration and light quantity on stomatal resistance in wheat leaves). Superimposed on the effects of these environmental factors are movements brought about by circadian rhythms and a possible variety of other, shorter-term rhythms.

The major role of stomata is to control entry of CO_2 into a leaf for photosynthesis and, at the same time, allow exit of water vapour which evaporates from the cell walls in contact with air spaces within the leaf. Some of the water lost serves for evaporative cooling of the leaf when exposed to high temperatures, but much is unnecessarily lost when the stomata are open. It has been estimated by Raschke (1979) that several hundred molecules of water are lost from a plant for each molecule of CO_2 taken up. Usually this means water must be conserved by a plant. The transpiration stream which is largely controlled by stomata also facilitates transport of ions from the roots. The uptake and transport of ions in the transpiration stream may, however, be less critical to the survival of the plant and need less fine control by stomata since active ion uptake and relocation of ions

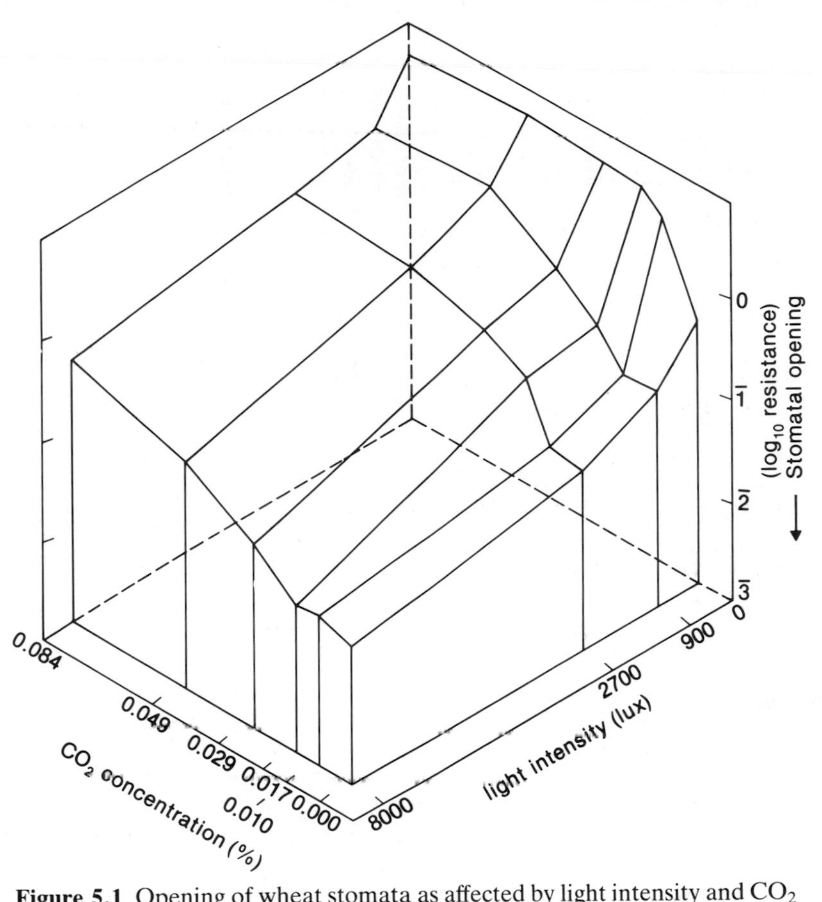

Figure 5.1 Opening of wheat stomata as affected by light intensity and CO_2 concentration. A higher value on the ordinate indicates stomtal closure (after Heath and Russell 1954).

within a plant will occur, to some extent, independently of transpiration. It seems, therefore, that it would be of advantage to a plant if stomata could respond to environmental stimuli by maximising photosynthesis while minimising water loss. In efforts to understand these opposing priorities Cowan (1977) and Raschke (1975) have applied a control system optimising gas exchange between a leaf and its environment. They envisage 'sensors' of environmental factors coupled with positive and negative feedback (closed loops) and feedforward (open loops) control systems which optimise the relationship between CO_2 uptake and water loss (Fig. 5.2). Raschke (1979) lucidly explains how this would be accomplished by the presence of two major feedback loops. One loop (b) is a regulator against a CO_2 deficiency caused by photosynthesis. When the

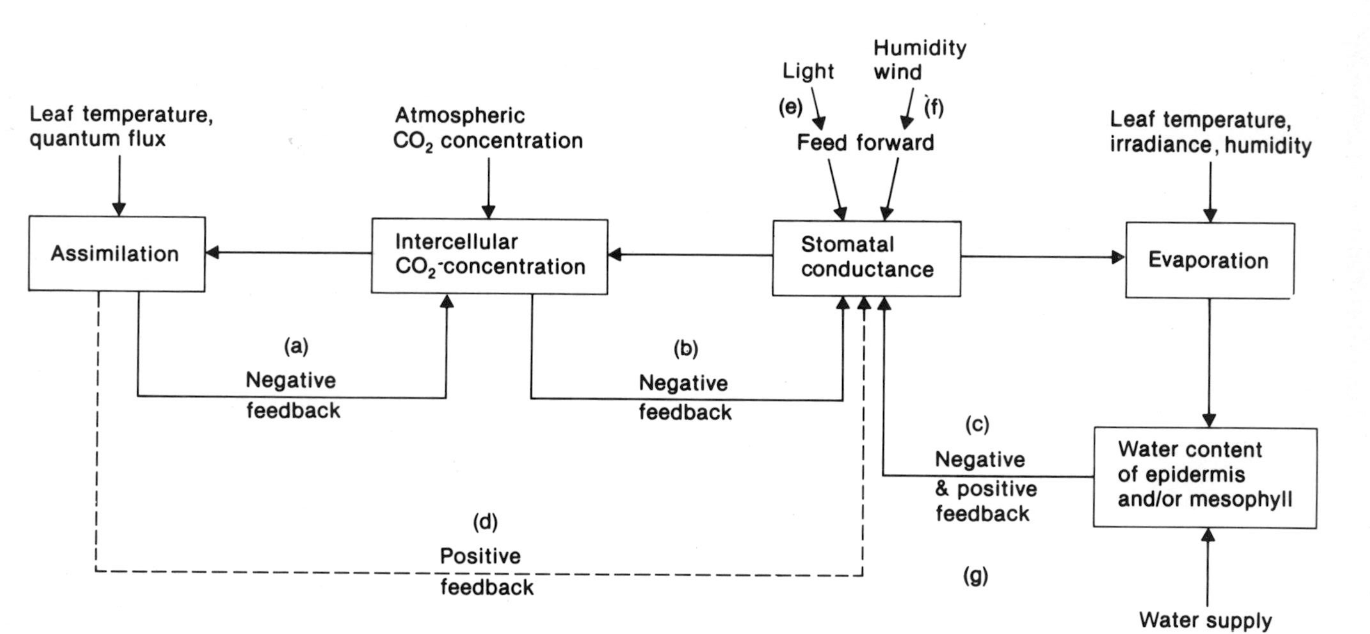

Figure 5.2 Control system utilising feedforward and feedback for optimising CO_2 uptake and water loss through stomata (Raschke 1979).

intercellular CO_2 concentrations becomes too low due to photosynthesis this is 'sensed' and a 'signal' (negative feedback) is sent to the stomata to open wider and allow more CO_2 to enter the leaf. The other loop (c) senses the water content of the leaves or epidermal tissue and, if the water content becomes too low, a 'signal' informs the stomata to close. The occasional observation of 'overshoot' when stomata open and oscillations of aperture which occur before equilibrium is reached is a typical response of such opposing loops.

Negative feedback of CO_2 (loops a, b) and water (loop c) alone will not optimise gas exchange between a leaf and the environment but would tend to keep intercellular CO_2 levels and tissue water levels constant (although this is dependent on the feedback gain and the time constants for the CO_2 and water loops of negative feedback to occur). This clearly does not happen in a leaf and, thus, some additional 'sensors' and 'signals' are required to optimise the gas exchange. These could be feedforward (open loops) due to a direct effect of light (e) or humidity and wind (f) and positive feedback from the photosynthetic processes (d) and water relations of the epidermal tissue and/or mesophyll (g).

It has also been suggested that stomata may optimise gas exchange by maintaining a constant ratio of the 'cost of water' against the 'benefit of carbon' (Cowan and Farquhar 1977).

It must be appreciated that the sensors and signals are largely hypothetical. However, abscisic acid (ABA) may be considered a 'signal' and cell turgor or water potential a 'sensor' in the negative feedback loop (c) since ABA accumulates in leaves under water stress and induces stomata to close. Such an analysis as that described above tends to over-simplify the complex biological system that exists. For example, the simple feedback loop involving ABA in reality describes a whole series of biological events. Thus, if our present understanding of the situation regarding the control of stomata by ABA is correct (but *see* §6.1) then a lowered water potential (or possibly turgor potential) triggers off ABA synthesis in the mesophyll chloroplast. The chloroplast must then become 'leaky' to ABA and, in a very directional and efficient transport system, the ABA must find its way to the epidermal layer where it eventually reaches the guard cells. Up to the point of reaching the cell neighbouring the guard cells, ABA may travel either symplastically or apoplastically. For the ABA to reach the plasmalemma of the guard cell it must diffuse through the common cell wall since plasmodesmata are very rare or absent between guard cells and neighbouring cells.

Wong *et al.* (1979) believe that some product or products of mesophyll photosynthesis can control stomatal movements via positive feedback (loop d); it is envisaged that such a metabolite which has some controlling influence on the rate of photosynthetic processes is transported to the guard cells. Thus, it is concluded by Wong *et al.* (1979) that stomatal aperture is a compromise between the need to conserve water and to maintain assimilation at a rate dependent on the capacity of the mesophyll to fix CO_2.

Over the years there have been disagreements about the effects of environmental factors on stomatal behaviour. For the first half of this century, for example, the consequences of changing CO_2 concentrations on stomata were not fully appreciated and the interpretation of many early experiments must remain dubious because of this. It was mainly the work of Heath and his colleagues which established the importance of controlling CO_2 concentrations surrounding leaves when stomatal behaviour was being monitored. Even in recent times the view that light and CO_2 have direct effects on stomatal movements has been occasionally questioned.

5.2 Light

Stomata of most plants open in response to light and close in response to darkness (Fig. 5.1). Stomata in CAM plants are an exception to this rule (*see* §5.3a). Stomata may also close during part of daylight and open during part of the night because of an endogenous effect (*see* §5.6) and certain species exhibit midday stomatal closure (*see* §5.4a).

Stomatal responses to light vary according to species, the age of the leaf or plant, the pretreatment of the leaf or plant and other accompanying environmental factors.

Relatively low levels of light (less than 1% of full sunlight) stimulate stomatal opening in leaves and the level needed to saturate the opening response is usually well below full sunlight. Although there is not complete agreement, many reports indicate that light levels of between 1 and 10 per cent of full sunlight are enough to saturate the opening response. Obviously the light saturation will depend on the pretreatment of the plant, the current environmental factors experienced by the plant and there may also be species differences.

In most species stomata on the abaxial (lower) leaf surface are more sensitive to light than the adaxial (upper) stomata: they open at lower light levels and have wider open stomata at all light levels than adaxial stomata. There are exceptions to this. For example, in peanut the stomatal resistance of adaxial stomata is less than that of the abaxial ones at the same light quantity.

Generally, closing is faster than opening in response to changing light quantitites. In some species (e.g. maize, cotton, sunflower), when light is incident on the adaxial leaf surface, the abaxial stomata open faster than the adaxial ones. However, if a leaf of sunflower is inverted so that light is incident on the abaxial surface then adaxial stomata open faster than the abaxial ones.

5.2a Indirect effects of light

Stomata are affected by light in a direct and an indirect way. An indirect effect of light can occur in the following manner. As the light quantity changes, so the photosynthetic rate of a leaf may change which, in turn, will change the intercellular CO_2 levels. Stomata will, therefore, respond to these changing internal levels of CO_2 within the leaf. This type of response may be particularly important in C_4 plants where the CO_2 gradient across stomata may be large, but it will be less so in C_3 plants where the CO_2 concentration drop across the stomata maybe 30 $\mu l\ l^{-1}$ or less. It has also been suggested that a 'messenger' substance, resulting from photosynthesis, may travel from the mesophyll cells to the guard cells so that leaf photosynthesis controls stomatal movements.

5.2 Direct effects of light

Light also has a direct effect on stomata. Early evidence for this view came from Heath and Russell (1954), who demonstrated a positive correlation in leaves of wheat between stomatal aperture and light quantity when intercellular CO_2 levels were kept constant. Wong *et al.* (1978) also concluded that light was the principal factor which tended to keep intercellular CO_2 concentrations constant in *Eucalyptus pauciflora* by altering stomatal conductances.

The response of stomata in epidermal strips and of guard cell protoplasts to light also provide convincing evidence that light has a direct effect on stomata. Additionally, most electrophysiological studies of guard cells indicated a direct effect of light on stomata since light hyperpolarises the guard cell plasmalemma (or possibly tonoplast), a phenomenon associated with increased ion fluxes into the guard cell.

Light can act as an energy source for cellular metabolic processes or as a signal to initiate some cellular process. Most species possess chlorophyll in their guard cells although the chloroplasts are usually poorly developed, but some photophosphorylation activity will occur. The ATP generated from such activity could be used for bringing about stomatal movements (*see* §7.3b). There is also evidence that another pigment (possibly a flavin or flavoprotein) is situated on or in the plasmalemma or tonoplast of the guard cell which, upon absorbing blue light, triggers off reactions resulting in stomatal opening.

Most investigators have found that blue light is two to twenty times more effective at causing stomatal opening than red light, while other wavelengths have little or no effect. Such responses have been observed both in intact leaves and in epidermal strips (Figs 5.3, 5.4). Guard cell protoplasts of onion also respond to blue light by swelling.

Differences in stomatal responses to blue and red light have been recorded between species. Blue light causes a rapid increase in transpiration rate before falling (overshoot) in species of Graminae and Cyperaceae, while in other species only a slow increase in transpiration rate is observed. Red light results in a slow increase in transpiration in all plants tested so far. These responses to red and blue light may be due, in part, to differences in stomatal anatomy, i.e. the grass-type stomata behaving differently to the elliptical stomata.

K^+ uptake (^{86}Rb used as a tracer) and concomitant stomatal opening in 'rolled' epidermal strips (epidermis is rolled with a glass rod which preferentially bursts the epidermal cells, leaving intact guard cells) of *Vicia faba* occur in blue light at low irradiance, while, at higher irradiance, there is also some stimulation of these processes in red light (Fig. 5.4). The maximum efficiency for K^+ uptake and stomatal opening is observed at 420 to 460 nm (Fig. 5.4).

The action spectrum for malate formation in sonicated epidermal strips (this disrupts the epidermal cells, but does not necessarily inactivate enzymes present, while leaving the guard cells intact) also indicates that blue light (peak at 433 nm) is more effective than red light (670–680 nm) (Ogawa *et al.* 1978). In the same study it was also shown that at low irradiance, blue light was much more effective with background red light than either red or blue light alone. It was postulated that two pigment systems were involved, one mediated by chlorophyll accounting for the red light and part of the blue light, and another pigment (possibly a flavin) accounting for the remaining blue light.

An action spectrum for maintenance of stomatal opening in epidermal strips of *Senecio odoris* has also been obtained and blue light was more than twice as effective as red light.

Superficially, the action spectra for photosynthesis and stomatal opening are similar. However, there is considerable evidence that another pigment system besides chlorophyll is involved in the control of stomatal movements. For example, blue light stimulates stomatal opening in *Paphiopedilum harrisseanum*, an orchid without guard cell chloroplasts. Blue light also induces the swelling of guard cell protoplasts of onion, a process enhanced in the presence of K^+ ions while other wavelengths of light had no effect. It has been suggested that stomatal movements may be phytochrome mediated but there is little support for this view, with most information indicating the involvement of a flavin or flavoprotein.

Blue light causes a decrease in the starch levels within guard cells, although the starch decrease may be an effect of stomatal opening brought about by blue light rather than a direct effect of the blue

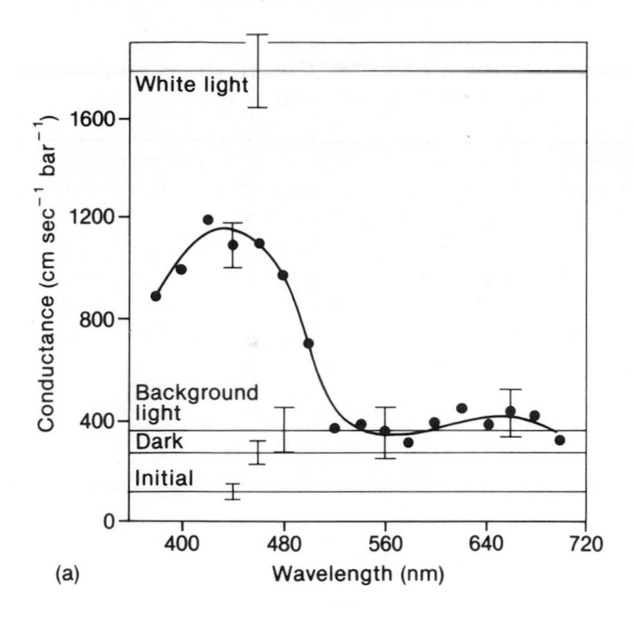

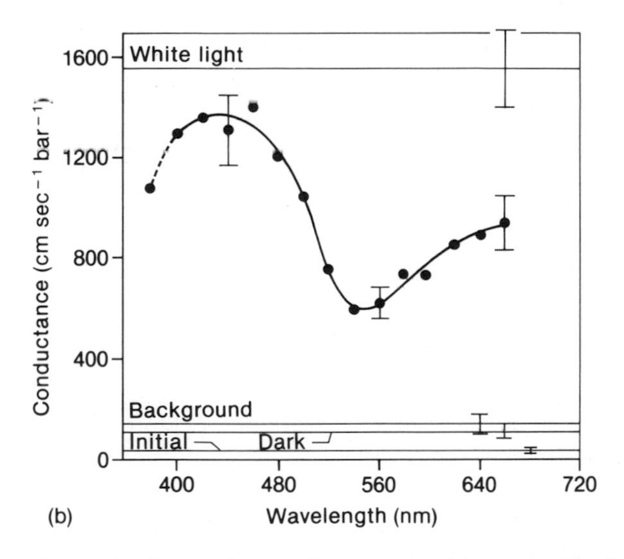

Figure 5.3 Spectral dependence at equal quantum flux density of stomatal opening in *Vicia faba* leaf discs. Exposure was for 3 hours under normal air and opening in the abaxial epidermis was measured with porometers. Vertical bars represent twice the standard error of the mean. (A) Low quantum flux density (7×10^{14} quanta cm^{-2} sec^{-1}). (B) High quantum flux density (38×10^{14} quanta cm^{-2} sec^{-1} except for 380 nm, which received 32.5×10^{14} quanta cm^{-2} sec^{-1}) (from Hsiao *et al.* 1973).

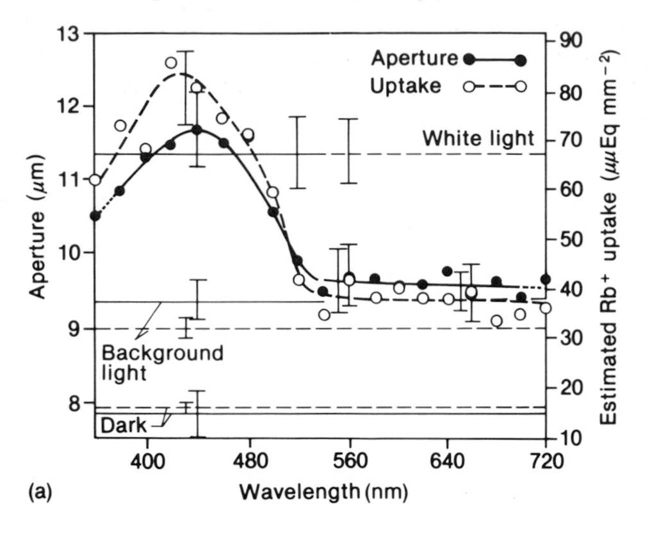

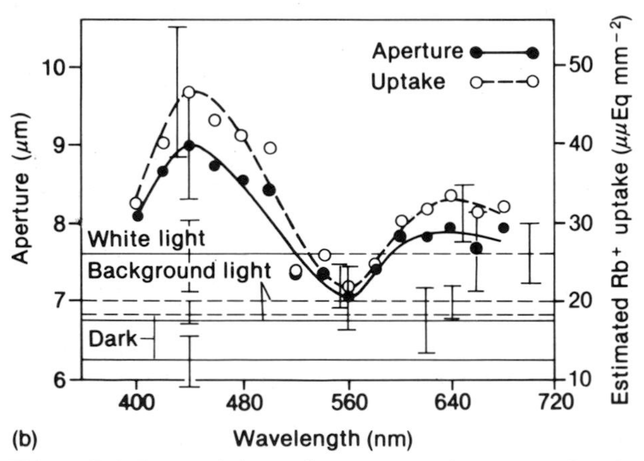

Figure 5.4 Spectral dependence at equal quantum flux density of Rb^+ uptake and stomatal opening in *Vicia faba* epidermal strips. Exposure was for 3 hours under CO_2-free air. Estimated uptake is based on mm^2 of epidermal area. Vertical bars represent twice the standard error of the mean for uptake or aperture of a particular treatment. (A) Low quantum flux density (7.8×10^{14} quanta cm^{-2} sec^{-1}) except for 360 and 720 nm, which received 5.9×10^{14} and 6.6×10^{14} quanta cm^{-2} sec^{-1}, respectively. (B) High quantum flux density (38×10^{14} quanta cm^{-2} sec^{-1}) (from Hsiao *et al.* 1973).

light. It is interesting to speculate that blue light may be affecting stomatal movements by altering the CO_2 levels within guard cells since it is known that blue light enhances respiration and PEP carboxylase activity.

5.3 Carbon dioxide

In general, as ambient CO_2 concentrations increase over the physiological range (atmospheric CO_2 concentration is about 320 $\mu l\ l^{-1}$) stomata close while, as CO_2 concentrations decrease over the same range, stomata open (Fig. 5.1). Thus, high CO_2 concentrations can close stomata, even in the light, while CO_2-free air can open stomata, even in the dark.

Stomata respond to the intracellular CO_2 concentration of the guard cells which is, in part, affected by the intercellular and ambient CO_2 concentrations. There is also evidence that high CO_2 concentrations affect membrane permeabilities and therefore CO_2 levels surrounding guard cells may also be of importance in controlling stomatal functioning.

Certain species under certain conditions (usually of saturating light intensities for stomatal opening) appear insensitive to changing CO_2 concentrations. The lack of response to CO_2 may occur when leaves have a high water potential and endogenous ABA is absent but the lack of response has been seen in a number of species with less than maximum leaf water potential.

Stomata of C_4 species also appear more sensitive to CO_2 than those of C_3 species and closure in C_4 species occurs at much lower CO_2 levels than observed in C_3 species. CO_2 levels below about 100 $\mu l\ l^{-1}$ may not cause further stomatal opening, at least in C_3 species, while in C_4 species stomata are apparently sensitive down to zero levels of CO_2.

Stomata of *Xanthium pennsylvanicum* and lettuce do not acclimatise to a particular CO_2 level, i.e. if plants are grown in high CO_2 concentrations for long periods of time the stomatal response to CO_2 is the same as in plants grown at 0 $\mu l\ l^{-1} CO_2$ (Mansfield and Jones 1970). This is possibly a phenomenon common to all plants.

The mechanism by which CO_2 regulates stomatal movements is not understood. Raschke (1979) has postulated that the level of CO_2 around the guard cells reflects malate levels within the cytoplasm of the guard cells. The cytoplasmic malate levels are considered to be determined by the rate at which malate is moved into the vacuole or out of the guard cell and its decarboxylation via malic enzyme and by its rate of synthesis via PEP carboxylase. Increased malic acid levels within the cytoplasm and concomitant lowered pH inhibit PEP carboxylase activity while malic enzyme activity increases with lowered pH. Thus, if stomata are open it is hypothesised that increased CO_2 levels increase the rate of malic acid synthesis in the cytoplasm more quickly than it can be disposed of and, as a result, malic acid levels will increase in the cytoplasm, pH levels will fall and stomata will close.

A weakness of the hypothesis is that such a system will result in fairly constant pH values and malate levels in the cytoplasm as a balance between malic acid synthesis and removal occurs. This is the basis of the postulated pH 'stat' in which cellular pH values are regulated and maintained fairly constant. In guard cells, when stomata close, vacuolar pH levels (and possibly cytoplasmic pH values, also) and malic acid levels decrease. Another dilemma regarding the hypothesis is that when stomata open malic acid levels increase in guard cells. A substrate for malic acid synthesis is CO_2 and, therefore, one would expect increased CO_2 supply would be needed for increased malic acid synthesis. However, increased CO_2 levels *close* stomata.

5.3a Stomatal movements in CAM plants

Stomata in CAM plants usually have substantial opening periods during the night and may close during the day. The exact pattern of stomatal movements, however, depends on the species and upon the pretreatment of the plant, particularly regarding diurnal temperature changes and the lengths of the photoperiod. In general, three patterns of stomatal movements have been observed (Neales 1975):

1. Stomata which open towards the end of the dark period and may remain open for the entire light period. These plants exhibit weak CAM.
2. Stomata which open for most of the dark period followed by increased opening immediately after the light period begins and then closure. Further opening occurs, again, at the end of the same light period. Such stomatal behaviour is found in plants exhibiting 'full' CAM.
3. Stomata which open during the dark period and close during the entire light period. These plants exhibit very strong CAM.

The behaviour of stomata in CAM plants is an adaptive mechanism evolved by plants which grow in hot and arid environments to conserve water. At night, when temperatures fall, stomata open to allow CO_2 to enter the plant. The CO_2 is 'fixed' and stored as malic acid. During the day when temperatures may be very high stomata close, preventing excessive water loss from the plant. The malic acid produced at night is essentially a CO_2 store and during the day it is decarboxylated, releasing CO_2 which is refixed via the Calvin cycle.

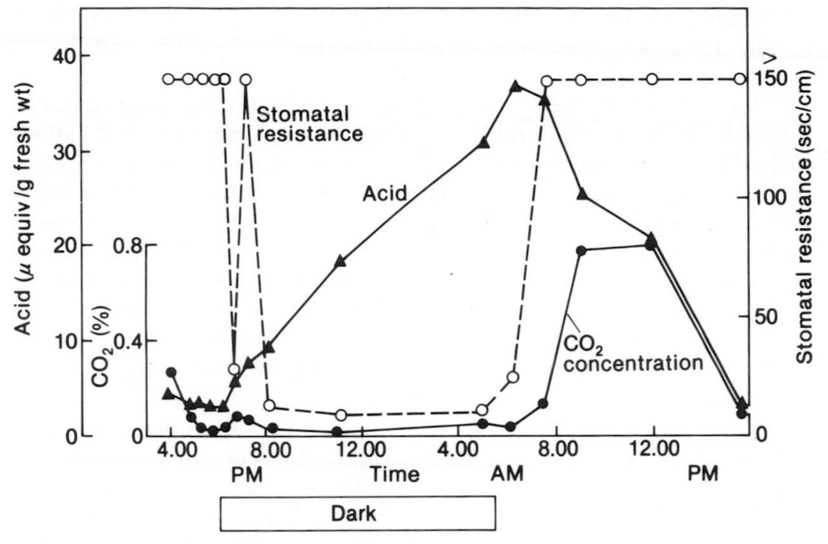

Figure 5.5 Diurnal changes in internal gas phase CO_2 concentration, stomatal resistance to diffusion of water vapour, and acid content in *Agave desertii* (from Cockburn et al. 1979).

The pattern of CO_2 exchange by CAM plants follows the course of stomatal movments, wider openings corresponding to increased CO_2 uptake. It was postulated that stomata controlled the CO_2 exchange rather than that the changing CO_2 concentrations within the leaves, which occur due to the carboxylation and decarboxylation phases of CAM, control stomatal movements. However, after removal of the epidermis CO_2 exchange remains much the same as when it is present. Moreover, a reasonably good positive correlation exists between intercellular CO_2 levels within the tissues and stomatal aperture (Fig. 5.5). It appears, therefore, that stomatal responses of CAM plants are primarily controlled by the changing intercellular CO_2 concentration in the following manner. When malate is being decarboxylated during the light period CO_2 levels within leaves will be high and stomata will close; when malate is being formed during the dark period CO_2 will be rapidly and efficiently fixed by PEP carboxylase, thereby lowering leaf CO_2 levels, and stomata will open.

5.4 Temperature

As the leaf temperature is raised, the metabolic activity within the guard cells and the leaf as a whole will increase, reach an optimum, and then decrease as more and more cell damage occurs. The effect of the increased metabolic activity within guard cells will be to stimulate opening.

There are also indirect affects of temperature on stomatal behaviour. It is possible that the temperature increases will affect internal CO_2 concentrations which, in turn, will affect stomatal movements. If respiration outpaces photosynthesis as the temperature increases, CO_2 levels will increase within the leaf which will tend to bring about closure. Also, an increase in leaf temperature will result in an increase of the vapour pressure (humidity) gradient between the leaf and the surrounding air which may cause stomatal closure either through a direct atmospheric humidity response (*see* §5.5) or by increasing transpirational water loss, resulting in plant water stress.

It is not surprising, therefore, that there is considerable variation in the temperature values obtained for optimal stomatal opening. For example, maximum opening in *Vicia faba* occurred at between 35 and 40°C (Stålfelt 1962) or even greater than 40°C for some species (Schulze *et al.* 1973) under some circumstances. Maximum opening in epidermal strips of the fern *Polypodium vulgare*, however, occurred at between 20 and 28°C depending on the pretreatment of the plants (Lösch 1977), while stomata in epidermal strips of *Commelina communis*, incubated in a medium containing K^+, reached maximum apertures at about 35°C (Willmer 1980).

Activation energies for stomatal opening have also been computed but equally variable values have been obtained and the importance and meaning of such numbers is debatable, particularly when intact leaves have been used and so many metabolic factors occurring in the leaf can indirectly affect stomata.

Increased temperature can overcome the closing effects of increased CO_2 levels and darkness. For example, increasing the temperature from 27 to 36°C resulted in stomatal opening in *Xanthium* leaves in the dark without removal of CO_2 (Mansfield 1965). Darwin (1898) also observed that closure occurred more slowly in the dark at high rather than low temperatures.

5.4a Mid-day closure

Stomata of some species such as coffee and onion show a distinct 'mid-day' closure particularly when growing under hot, sunny conditions (Fig. 5.6). Initially the closure was attributed to water stress brought about by the high temperatures which occurred about the middle of the day (Loftfield 1921). However, later workers considered it was a CO_2 effect. As temperatures increase, respiration outpaces photosynthesis and high CO_2 levels accumulate in the leaves which close stomata. Evidence for this view came from Meidner and

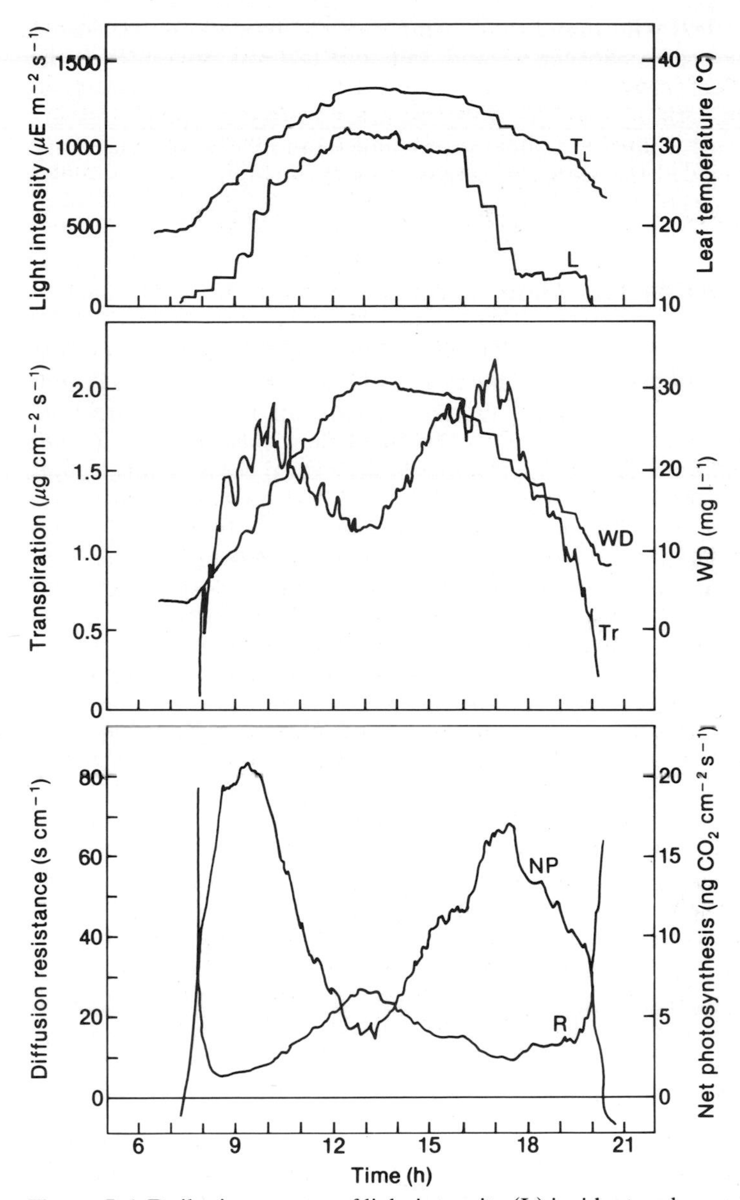

Figure 5.6 Daily time course of light intensity (L) incident on leaves in the gas exchange cuvette, leaf temperature (T_L), transpiration rate (Tr), water vapour concentration difference (WD), leaf diffusion resistance (R) and net photosynthesis rate (NP) measured for leaves of well watered potted *Arbutus unedo* plants under simulated natural conditions in an environmental chamber (from Tenhunen *et al.* 1980).

Heath (1959) who found that, using onion leaves, stomatal closure normally occurs between 25 and 35°C but this was prevented if the central leaf cavity was flushed with CO_2-free air. However, as indicated earlier, it is difficult to maintain constant vapour pressure deficits with changing temperatures and humidity factors may have still influenced the stomatal activity to some extent.

5.5 Humidity and plant water status

Stomata respond to bulk leaf water potential by closing if the leaf water content becomes too low and opening as the leaf water content increases. In some species stomata remain open until a threshold level of leaf water deficit is reached and then close rapidly. Under field conditions leaf water potentials may drop to very low values before the stomata close.

The stomata of some species also respond to changes in atmospheric humidity (atmospheric water vapour pressure deficit) by opening as humidity increases and closing as the atmosphere becomes drier (Fig. 5.7). Whether these responses to humidity are due to direct or indirect effects on stomata is not clear and nor are the effects of the interaction of plant water balance and atmospheric humidity on stomatal movements completely understood.

Indirect effects due to hydropassive movements can result from changing humidity. For example, decreased humidity will result in a steeper diffusion gradient from inside the leaf to the atmosphere, resulting in an increased rate of water vapour diffusion out of the leaf. This could cause a decreased bulk leaf water potential with ultimate stomatal closure or, alternatively, if leaf water loss is initially restricted to the epidermal layer, epidermal cell turgor may be lowered relative to the guard cells with concomitant stomatal opening. If the leaf becomes sufficiently water-stressed stomata will then close.

A hydroactive mechanism and direct effect of humidity on guard cell metabolism has been postulated but no definite mechanism is known. The main evidence for this view is that in some species dry air can cause stomatal closure without measureable changes in bulk leaf water potential (Fig. 5.7). However, bulk leaf water potential may not be the best indicator of water availability to the guard cells especially in those species with a low water conductivity between the epidermal layer and the mesophyll.

Guard cell metabolism could be affected in an indirect way by humidity in the following manner. As leaf water stress increases due to increased water vapour loss from the leaf into dry air, ABA levels will increase and this will bring about stomatal closure. Since our

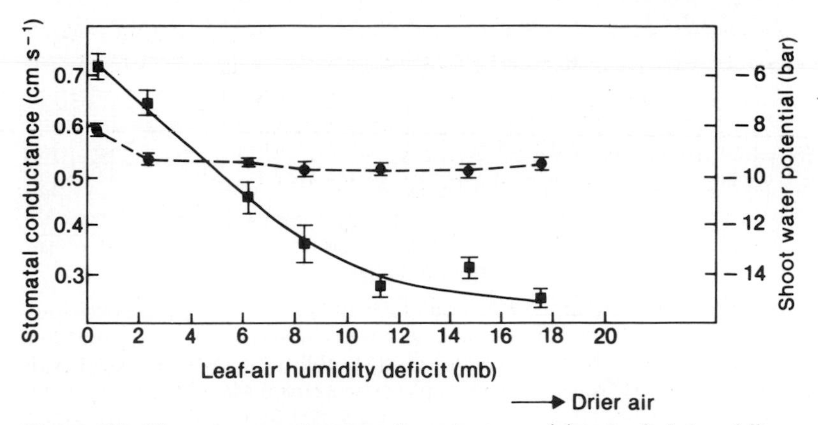

Figure 5.7 The response of stomatal conductance (■) to leaf-air humidity deficit in potted Sitka spruce seedlings. Note that the response is observed at constant shoot water potential (●) (after Watts and Neilson 1978).

present knowledge indicates that ABA is not synthesised in guard cell chloroplasts but in chloroplasts of mesophyll cells the humidity effect on stomata may be considered an indirect one.

There is some evidence that leaf water stress increases stomatal sensitivity to humidity changes while high light intensity reduces stomatal sensitivity to humidity.

There are also species differences in stomatal responses to humidity although this has not been related to a particular type of leaf anatomy. Nevertheless there may be an ecological advantage to those plants which are sensitive to changing atmospheric humidity.

5.5a Peristomatal transpiration

At one time it was considered that ectodesmata existed as cytoplasmic strands, extending from the guard and epidermal cells to the outer cuticular layers of the cell walls. It is now known that such structures do not exist but that there are very permeable regions in the outer walls and cuticle of the epidermal layer, particularly over guard cells. Water loss through these regions is known as peristomatal transpiration and, under certain circumstances, the loss of water can greatly increase cuticular transpiration. Peristomatal water loss may also account for occasional observations of a lack of a positive correlation between stomatal aperture and transpiration rate.

5.5b Iwanoff effect

When a leaf is excised, Iwanoff (1928) observed that there was a temporary increase in stomatal opening before the stomata per-

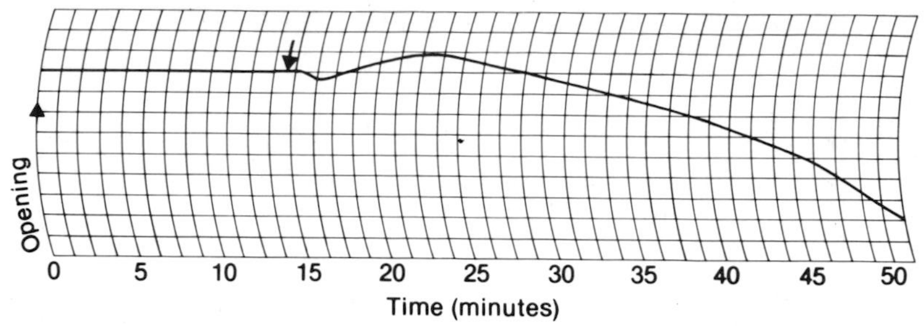

Figure 5.8 A porometer trace of stomatal behaviour following leaf excision. The stomata were open in light and leaf was excised at the time indicated by the arrow. Note the initial small closure, followed by a period of wide opening before the final closing response occurs (from Mansfield 1971).

manently closed. Later studies observed that there was a momentary stomatal closing effect before opening occurred followed by the closure (Fig. 5.8). These observations have been explained in the following manner. When the leaf is excised the tension of the water columns in the xylem vessels is released. This results initially in a transitory pressure on the guard cells with concomitant slight closure as water first hydrates the surrounding epidermal and subsidiary cells, but then the guard cells successfully compete for the water with resulting stomatal opening. Ultimately, due to transpirational water loss, water becomes limiting to the whole leaf and the stomata close.

5.6 Circadian rhythms and shorter-term oscillations in stomatal movements

Plants and animals exhibit rhythmic behaviour in their biochemical and physiological activity, usually as a result of the diurnal pattern of day and night with associated temperature changes which is imprinted into the organism. A diurnal rhythm is one having a periodicity of precisely 24 hours. However, almost all free-running rhythms in animals and plants have a periodicity of between 21 and 28 hours, but not exactly 24 hours, and such rhythms are known as circadian (*circa* = about; *diem* =day). Circadian rhythms and many other longer or shorter term rhythms are endogenous, i.e. a rhythm programmed from within the organism as opposed to one brought about by exogenous stimuli.

To demonstrate that a rhythm is endogenous five features should be observed (Pittendrigh 1954): (1) the period of the rhythm must be about 24 hours but not precisely 24 hours; (2) the rhythm should persist under constant environmental conditions (in practice this is

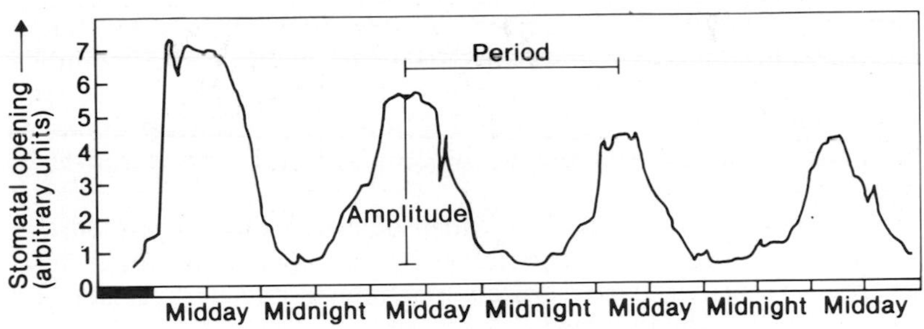

Figure 5.9 Stomatal behaviour in *Tradescantia virginiana* leaves in continuous light (1,500 lux) at 20°C (adapted from Martin and Meidner 1971).

difficult to achieve – geophysical factors, for example, are difficult to control); (3) it must be possible to phase-shift the rhythm; (4) the rhythm should be initiated by a single stimulus (the rhythm is considered to be inherited and the period of the rhythm is not acquired from exposure to cyclical variation in the environment); (5) the rhythm is arrested by anoxia.

Circadian rhythms in stomatal movements were observed long ago (Darwin 1898; Lloyd 1908), but even today our understanding of them is not clear. It is not known to what extent they control stomata and nor is their value to the plant understood.

When a plant is kept in continuous light the stomata open and close with a period of about 24 hours (Fig. 5.9). There is a damping or gradual decrease in the amplitude of the free-running or natural period of the rhythm with time. Although this is highly suggestive of an endogenous rhythm it is difficult to establish with certainty that the rhythm was endogenous unless intercellular CO_2 concentrations are controlled. If there was an intracellular circadian rhythm of CO_2 production, as in CAM plants, (Wilkins 1959), this alone could effect stomatal movements, giving an appearance that the stomata themselves were controlling this rhythm. However, if the guard cells produced the CO_2 then the stomatal movements would be considered circadian.

The rhythm in continuous light can be phase shifted so that a 6-hour dark period given during the natural 'light' phase causes an inversion of the original pattern.

Stomata also exhibit rhythms in continuous dark although this is less apparent than in continuous light. In continuous dark only one cycle of opening has been observed in *Xanthium pennsylvanicum*, *Tradescantia virginiana* and *Vicia faba*, although in the latter species Stålfelt had observed two cycles of opening (Martin and Stevens 1979). The opening cycle in *T. virginiana* in continuous dark can be phase-shifted when a period of light occurs during the natural 'dark' phase but there is little or no phase shift when a period of light occurs during the natural 'light' period. A low intensity light-treatment (10–1,500 lux) of *Xanthium pennsylvanicum* plants at the beginning of the dark treatment causes a phase shift, or, effectively, delayed opening in the dark (Mansfield and Heath 1963). Also, the timing of both light and dark treatments affects the timing and duration of opening in darkness.

The low light intensity effect which delays opening in the dark is temperature dependent; no delay occurs in light of 10 lux at 15°C whereas, at higher temperatures, opening in the dark is delayed (Mansfield and Heath 1964).

Circadian rhythms in 'osmotic surplus' of guard cells (the difference in osmotic potential between guard cells and neighbouring epidermal cells) in *V. faba* plants growing in continuous darkness and in protoplasmic streaming in guard cells of *V. faba* in continuous darkness have also been observed.

5.6a Night opening of stomata

The term night opening normally refers to stomatal opening which occurs before sunrise; stomata may also remain open after sunset. It represents part of an endogenous rhythm which overlaps with other opening stimuli during daylight and night hours. Stomatal opening also occurs at night in CAM plants but this appears to be largely controlled by intercellular CO_2 concentrations. It is likely that there is also some endogenous control of apertures by the guard cells in CAM plants.

The basic mechanism which times the endogenous rhythm (biological clock) of stomatal movements or any other endogenous rhythmic process has not been identified. Rhythmic change in the levels of some metabolite or energy source may be key controls but, in turn, some basic oscillating process must control the cycling of the metabolite levels. It has been suggested that the cellular energy supply is concerned in the control of endogenous rhythms since anoxia, in some cases, inhibits the rhythm. The lack of certain metabolites or energy source may account for the unsustained rhythm of stomatal opening when leaves are kept in continuous darkness.

The function of circadian rhythms in the control of stomatal aperture is not understood. They may be of some adaptive significance and in some organisms they have been shown to be of survival value.

Stomata also exhibit a variety of shorter term oscillations of aperture. In general these are brought about by sudden changes of environmental factors which effect the water balance of the plant.

In some species oscillations in stomatal aperture with a period of about 20 to 40 minutes (they may be up to $2\frac{1}{2}$ hours in the dark) are caused by time lags in the rehydration of leaves. The time lags, in turn, are caused by a combination of the hydraulic resistances and hydraulic capacitances in the plant. The sequence of events can be summarised as follows: If the transpiration rate is suddenly increased (e.g. by increased irradiance) the rate of water uptake can lag behind the transpiration rate. This results in the leaf becoming water stressed with resultant hydropassive stomatal closure (in some species there may be a transient stomatal opening – see the Iwanoff effect, section 5.5b). The leaf rehydrates and the stomata reopen. This sequence of events is then repeated until, finally, an equilibrium situation is reached and the stomatal oscillations damp out.

Fluctuations of transpiration (with a period of about 1–10 minutes) may occur *without* concomitant changes in stomatal aperture. This has been explained in much the same way as described above for stomatal oscillations except that the rehydration and dehydration cycles of the symplast are affected much less than the apoplast. Thus the turgor relations of the cells are not greatly affected and therefore stomatal movements do not occur.

5.7 Gaseous environmental pollutants and stomata

Major gaseous environmental pollutants which may effect plants are SO_2, oxides of nitrogen, ozone and certain hydrocarbons. These substances are produced by industry, often in very localised areas, by combustion engines and domestic heating systems in their exhaust gases, and some pollutants may be formed from photochemical reactions in the atmosphere. The latter air pollution is known as 'photochemical smog'.

Two areas of major interest are; (1) the effect of the pollutant on stomatal behaviour and the method of entry of the pollutant into leaves; and (2) the effect of the pollutant on the physiology and growth of the plant. The latter aspect will be neglected in the discourse here.

Sulphur dioxide has received most attention but reports on the effects of this pollutant on stomatal behaviour are conflicting. Many researchers find that relatively low concentrations of SO_2 (50–2,860 $\mu g\ m^{-3}$, equivalent to 0.02–1 $\mu l\ l^{-1}$) increase stomatal opening and transpiration rates while reducing photosynthesis. Other investigators observe either no effect of low SO_2 concentrations on transpiration or that stomatal closure is brought about by relatively high SO_2 concentrations (1–2 $\mu l\ l^{-1}$). To account for the differences in stomatal behaviour in some species to high and low SO_2 concentrations Black and Black (1979) suggested the following. Associated with wider stomatal opening in *Vicia faba* leaves exposed to SO_2 (from 50–200 $\mu g\ m^{-3}$) was increased epidermal cell damage which could result in hydropassive opening, while at higher SO_2 concentrations (500 $\mu g\ m^{-3}$ and above) guard cells were damaged which would result in stomatal closure.

Environmental factors also appear to have an important control in the sensitivity of stomata to SO_2. For example, the effects of SO_2 on stomata in some species are dependent on atmospheric humidity (Mansfield and Majernick 1970; Black and Black 1979), while growth of the grass, *Phleum pratense*, is more sensitive, i.e. growth was reduced, to SO_2 pollution in conditions of low irradiance and short days (Davies 1980).

The extent of leaf damage appears to be related to stomata aperture. Generally, symptoms of leaf injury are not observed if stomata are closed. Leaf injury may also be related to the initial degree of stomatal opening and the subsequent rate of closure upon contact with SO_2. For example, Bonte and Longuet (1975) found that increased atmospheric humidity slowed down stomatal closure in *Pelargonium zonale* (= *Pelargonium* × *hortorum*) leaves which, in the presence of SO_2, resulted in increased leaf necrosis compared to leaves exposed to SO_2, at lower atmospheric humidites.

Some species are much more sensitive to SO_2 damage than others for reasons not understood. Peanut and tomato plants, for example, appear more resistant to the effects of SO_2 than radish or spinach.

Photochemical smogs occur over large cities wherever high levels of oxides of nitrogen and hydrocarbons accumulate in the atmosphere. Photochemical reactions occur which result in further products such as ozone and peroxyacetyl nitrate being formed. Photochemical smog induces stomatal closure and, in general, increased stomatal aperture and damage to plants are positively correlated.

The effects of individual components of photochemical smogs on stomatal behaviour and leaf damage have been studied. Ozone has been observed to usually close stomata although it is not clear whether this is due to its direct affect on stomata or due to its inhibition of photosynthesis resulting in increased CO_2 levels within the leaf. The concensus of opinion is that if stomata are closed there is much less damage to foliage. Also, some plants are more resistant to the effects of ozone than others. This increased resistance has been correlated with a lower stomatal frequency which can, result in slower penetration of the pollutant.

Exposure of foliage to a few hundred parts per million (0.01–0.1 $\mu l\ l^{-1}$) of ozone for a few hours can result in damage to some species. One symptom of ozone damage is decreased chlorophyll levels within leaves.

Less is known about the effects of NO_2 although most investigators observe decreased transpiration rates and photosynthesis, suggesting that stomatal apertures are also decreased. Nitrous oxide (NO), however, has no effect on stomata.

To summarise, some plants seem more resistant to a pollutant than others and, generally, there is less damage to a plant if the stomata are closed. Of course, closed stomata may not be a total barrier to the pollutants which may enter leaves by penetrating the cuticle. Thus, the rate of entry of a pollutant and, therefore, the external concentration of the pollutant and the exposure period are important in determining the extent of damage to a plant.

The long-term effects of exposure to a pollutant on stomatal behaviour and plant damage are also likely to be considerably different to shorter-term exposures. Environmental factors may also play a large role in determining the extent of the damage caused by the pollutants. Exposure of plants to different combinations of pollutants may also result in markedly different and possibly more drastic effects than exposure of plants to individual pollutants.

5.8 Plant pathogens and stomata

5.8a Entry of pathogens into leaves via stomata

A number of plant pathogens, including some of economic importance, gain entry into foliage either solely via stomata or as an alternative to direct penetration. Such pathogens include parasitic bacteria, certain downy mildews, aecidiospores and uredospores of *Puccinia* spp. (rusts) and several *Fungi Imperfectii.*

Entry through stomata by fungal pathogens may be grouped into three types. One type is that observed when cereal leaves are infected by uredospores of *Puccinia* spp. Upon germination of a uredospore the germ tube shows directional growth towards a stoma. Upon reaching a stoma the cytoplasm of the germ tube accumulates at the tip which swells to form an appressorium within the stomatal pore. A cross-wall forms, cutting off the empty germ tube, and from the appressorium a wedge-shaped hypha grows through the pore. Inside the substomatal cavity the hypha swells, receiving the contents of the appressorium.

Another type of pathogen entry is found in *Fulvia* (formerly *Cladosporium*) *fulvum* in which spores germinate on the leaf surface in a water film or in moist air. Growth of the germ tube appears to be haphazard and they often grow over stomata. When penetration of a stoma occurs a lateral hypha forms which grows through the pore into the substomatal cavity.

A third type of stomatal penetration by a fungal pathogen is exemplified by that found in *Plasmopara viticola*. Motile zoospores swim in a water film on the leaf surface and, upon nearing a stoma, are attracted to them by some stimulus. The zoospores then settle over the stomata and encyst. A hypha emerges from the cyst and grows through the stomatal pore into the substomatal cavity.

The nature of the stimulus which attracts zoospores and germ tubes to stomata and causes encystment or appressorium formation has received considerable attention. A number of explanations have been forwarded including attraction of the pathogens along chemical or water vapour gradients or that the pathogen is directed by the surface topography of the leaf (a thigmotropic response).

Royle (1976) has suggested that zoospores of *Pseudoperonospora humuli*, which tend to select open stomata, are attracted by two stimuli. One stimulus is surface topography since zoospores on perspex replicas of leaf surfaces tend to settle on open stomata rather than closed ones. The other stimulus was a chemical one and it was suggested to 'involve photosynthesis'. It is enticing, therefore, to suggest that a CO_2 gradient attracts zoospores to stomata. However, the evidence indicating that CO_2 concentrations regulate entry into stomata is not strong although high, non-physiological, levels of CO_2 have been observed to suppress penetration by some pathogens including *Puccinia graminis*.

Another intriguing question is how surface topography can direct germ tubes to stomata. A partial answer to this question is that the germ tubes are directed to grow between rows of wax crystals which form a regular lattice on some leaf surfaces. This may also explain why a spore from an obligate parasite which germinates on a plant other than its normal host may not develop a proper infection structure. In other words, the germ tube is unable to adapt to different dimensions of the regular lattice of wax crystals of the host leaf surface and becomes disorientated.

Attempts have also been made to find a relationship between stomatal frequency and their arrangement and structure and resistance to pathogen entry. Although there are conflicting reports it is likely that stomatal frequency and structure will influence the rate and numbers of pathogen infection. Some pathogens enter the leaf through open stomata while others can force an entry even through closed stomata. In *Citrus nobilis*, a species of orange, the stomatal anatomy regulates pathogen entry. The upper cuticular lips of the guard cells are very pronounced and these, apparently, prevent entry of *Pseudomonas citri* (a bacterial canker) and *C. nobilis* is, therefore, resistant. In *C. grandis*, however, the cuticular lips are not so prominent and the bacterial pathogens can more readily enter the stomatal pores, making this species of orange much more susceptible.

5.8b Effects of pathogens on stomatal behaviour

When pathogens infect plants, stomata may be affected in a variety of ways which may be reflected in different patterns of transpiration and

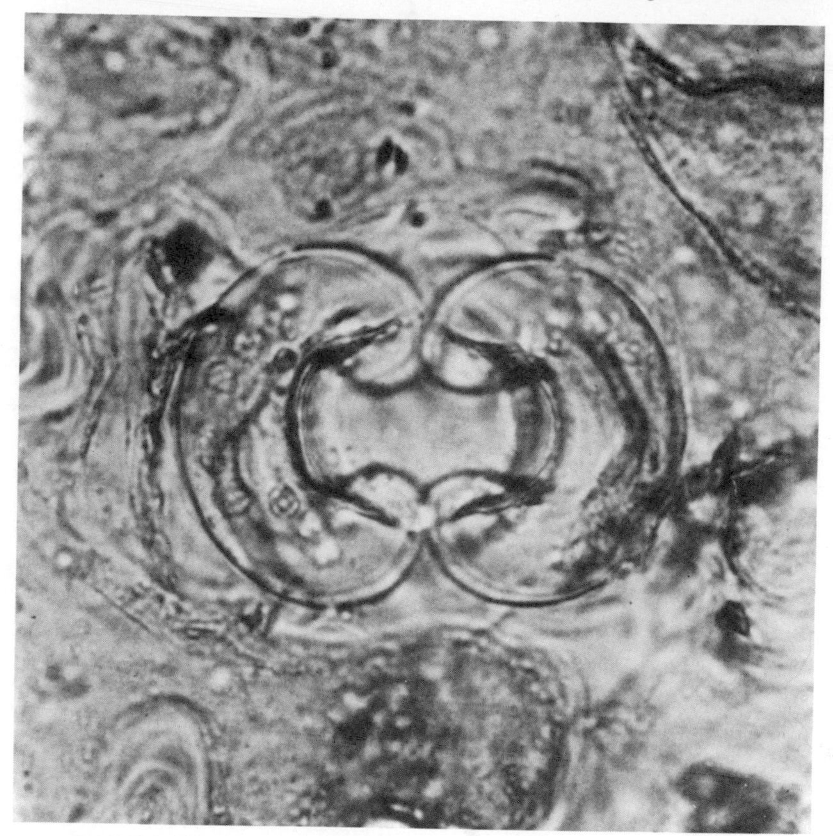

Figure 5.10 Unnaturally wide stomatal opening in *Vicia faba* as a result of infection by *Botrytis fabae* (Mansfield, J. and Willmer, C. M., unpublished).

photosynthesis. The host/pathogen interaction is often complex and there may be direct and indirect effects of the pathogen on stomata. The direct effects may be due to substances produced by the host (phytoalexins), or substances produced by the pathogen (toxins, enzymes) which affects the stomata in some way. An example of how stomata are affected in an indirect way by a pathogen is found in the vascular wilts (e.g. *Fusarium oxysporum lycopersici*) which can cause rapid wilting of leaves and eventual death of the host. The major cause of these symptoms is that the vessels become plugged with fungal hyphae, cell debris from damaged walls or tyloses and the water supply to the leaves becomes restricted. As a result, stomatal apertures may fluctuate even before visual symptoms of infection appear though ultimately stomata will close through lack of water supply. In only a very few cases is there evidence that membrane-damaging toxins are involved in the vascular wilt syndrome.

In general, when pathogens infect leaves stomata lose their control of gas exchange between the leaf and the environment. In some cases stomata in infected leaves close or eventually fail to open widely in light which will not only reduce transpiration but also photosynthesis and growth of the plant. In other cases stomata remain wide open even in the dark and there will be uncontrollable loss of water. Both opening and closing responses may be brought about by production of phytoalexins by the host or phytotoxins by the pathogen which affect guard cell processes directly (§6.5 and 6.6). Wide stomatal opening and uncontrollable loss of water may also be due to destruction and digestion of the cuticle and cell walls of the epidermal layer by enzymes secreted by the pathogen. Figure 5.10 shows extremely wide open stomata in leaves of *Vicia faba* infected with *Botrytis fabae*. The state of the stoma is believed to be due to enzymes liberated by the pathogen which have partially digested the walls of the guard cells thereby allowing an unnatural extension of the guard cells. Phytotoxins, extracellular enzymes and phytoalexins may also affect membrane permeability (*see* §6.5 to 6.6).

More detailed information on the effects of pathogens on stomatal behaviour may be found in a report by Duniway (1976).

Further reading and references

Black, C. R. and Black, V. J. (1979). The effects of low concentrations of sulphur dioxide on stomatal conductance and epidermal cell survival in field bean (*Vicia faba* L.), *J. Exp. Bot.*, **30**, 291–8.

Bonte, J. and Longuet, P. (1975). Interrelations entre la pollution par le dioxyde de soufre et le mouvement des stomates chez le *Pelargonium* × *hortorum*: effets de l'humidité relative et de la teneur en gaz carbonique de l'air, *Physiol. Vég.*, **13**, 527–37.

Cockburn, W., Ting, I. P. and Sternberg, L. O. (1979). Relationships between stomatal behaviour and internal carbon dioxide concentration in crassulacean acid metabolism plants, *Plant Physiol.*, **63**, 1029–32.

Cowan, I. R. (1977). Stomatal behaviour and environment, *Adv. Bot. Res.*, **4**, 117–228.

Cowan, I. R. and Farquar, G. D. (1977). Stomatal function in relation to leaf metabolism and environment, in Integration of Activity in the Higher Plant, *Soc. Exp. Biol. Symp.*, **31**, 471–505.

Darwin, F. (1898). Observations on stomata, *Phil. Trans. R. Soc. B.*, **190**, 531–621.

Davies, T. (1980). Grasses more sensitive to SO_2 pollution in conditions of low irradiance and short days, *Nature*, **284**, 483–5.

Duniway, J. M. (1976) Water status and imbalance, in Physiological Plant

Pathology (Eds R. Heilefuss and P. H. Williams) *Encyclopedia of Plant Physiology*. New Series, vol. 4. Springer-Verlag, Berlin.

Heath, O. V. S. and Russell, J. (1954). Studies in stomatal behaviour. Vl. An investigation of the light responses of wheat stomata with the attempted elimination of control by the mesophyll, *J. Exp. Bot.*, **5**, 1–15.

Hopmans, P. A. M. (1971). Rhythms in stomatal opening of bean leaves, *Meded. Landbouwhogeschool Wageningen*, **71**, 1–86.

Hsiao, T. C., Allaway, W. G. and Evans, L. Y. (1973). Action spectra for guard cell Rb^+ uptake and stomatal opening in *Vicia faba*, *Plant Physiol.*, **51**, 82–8.

Iwanoff, L. (1928). Zur Methodik der Transpirations-bestimmung am Standort, *Ber. Deutsch. Bot. Ges.*, **46**, 306–10.

Lloyd, F. E. (1908). The physiology of stomata, *Publ. Carnegie Inst., Wash.* Publ. No. 82, 1–142.

Loftfield, J. V. G. (1921). The behaviour of stomata, *Publ. Carnegie Inst., Wash*. Publ. No. 314, 1–104.

Lösch, R. (1977). Responses of stomata to environmental factors–experiments with isolated epidermal strips of *Polypodium vulgare*. 1. Temperature and humidity, *Oecologia*, **29**, 85–97.

Mansfield, T. A. (1965). Stomatal opening in high temperature in darkness, *J. Exp. Bot.*, **16**, 721–31.

Mansfield, T. A. (1971). Stomata: versatile sensory devices but difficult experimental subjects, *J. Biol. Educ.*, **5**, 115–23.

Mansfield, T. A. and Heath, O. V. S. (1963). Studies in stomatal behaviour. IX. Photoperiodic effects on rhythmic phenomena in *Xanthium pennsylvanicum*, *J. Exp. Bot.*, **15**, 334–52.

Mansfield, T. A., and Heath, O. V. S. (1964). Studies in stomatal behaviour. X. An investigation of responses to low intensity illumination and temperature in *Xanthium pennsylvanicum*, *J. Exp. Bot.*, **15**, 114–24.

Mansfield, T. A. and Jones, R. J. (1970). Increases in the diffusion resistances of leaves in a carbon dioxide-enriched atmosphere, *J. Exp. Bot.*, **21**, 951–8.

Mansfield, T. A. and Majernick, O. (1970). Can stomata play a part in protecting plants against air pollutants?, Environ. Pollut., 1, 149–54.

Mansfield, T. A. and Meidner, H. (1966). Stomatal opening in light of different wavelengths: effects of blue light independent of carbon dioxide concentration, *J. Exp. Bot.*, **17**, 510–21.

Martin, E. S. and Meidner, H. (1971). Endogenous stomatal movements in *Tradescantia virginiana*, *New Phytol.*, **70**, 923–8.

Martin, E. S. and Stevens, R. A. (1979). Circadian rhythms in stomatal movements. Sen, D. N. (ed), in *Structure, Function and Ecology of Stomata*. Bishen Singh and Mahendra Pal Singh, Dehra Dun, India.

Meidner, H. and Heath, O. V. S. (1959). Stomatal responses to temperature and carbon dioxide concentration in *Allium cepa* L. and their relevance to midday closure, *J. Exp. Bot.*, **10**, 206–19.

Neales, T. F. (1975). The gas exchange patterns of CAM plants in *The Environmental and Biological Control of Photosynthesis* (Ed. R. Marcelle). Junk, The Hague.

Ogawa, T., Ishikawa, H., Shimada, K. and Shibata, K. (1978). Synergistic action of red and blue light and action spectra for malate formation in guard cells of *Vicia faba* L.,*Planta*, **142**, 61–5.
Pittendrigh, L. S. (1954). On the temperature independence in the clock system controlling emergence time in *Drosophila*, *Proc. Natl. Acad. Sci., USA*, **40**, 1018.
Raschke, K. (1975). Stomatal action, *Ann. Rev. Plant Physiol.*, **26**, 309–40.
Raschke, K. (1979). Movements of Stomata, in Haupt, W. and Feinlieb, M. E. (eds), *Encyclopedia of Plant Physiology*. vol. 7, Physiology of movements. Springer-Verlag, Berlin.
Royle, D. J. (1976). Structural features of resistance to plant-diseases, in Friend, J. and Threlfall, D. R. (eds), Biochemical Aspects of Plant-Parasite Relationships, *Phytochem. Soc. Symp. Ser. No. 13*. Academic Press, London.
Schulze, E. D., Lange, O. L., Kappen, L., Buschbom, U. annd Evanari, M. (1973). Stomatal responses to changes in temperature at increasing water-stress, *Planta*, **110**, 29–42.
Stålfelt, M. G. (1962). The effect of temperature on opening of stomatal cells, *Physiol. Plant.*, **10**, 752–93.
Tenhunen, J. D., Lange, O. L., Braun, M., Meyer, A. Lösch, R. and Pereira, J. S. (1980). Midday stomatal closure in *Arbutus unedo* leaves in a natural macchia and under simulated habitat conditions in an environmental chamber, *Oecologia*, **47**, 365–7.
Watts, W. R. and Neilson, R. E. (1978). Photosynthesis in Sitka spruce (*Picea sitchensis* [Bong.] Carr.) VIII. Measurements of stomatal conductance and $^{14}CO_2$ uptake in controlled environments, *J. Appl. Ecol.*, **15**, 245–55.
Wilkins, M. B. (1959). An endogenous rhythm in the rate of carbon dioxide output of *Bryophyllum*. 1. Some preliminary experiments, *J. Exp. Bot.*, **10**, 377–90.
Wilkins, M. B. (1969). Circadian rhythms in plants, in Wilkins, M. B. (ed.) *Physiology of Plant Growth and Development*. McGraw Hill, London.
Willmer, C. M. (1980). Some characteristics of phosphoenolpyruvate carboxylase activity from leaf epidermal tissue in relation to stomatal functioning, *New Phytol.*, **84**, 593–602.
Wong, S. C., Cowan, I. R. and Farquhar, G. D. (1978). Leaf conductance in relation to assimilation in *Eucalyptus panciflora* Sieb. ex Spreng. Influence of irradiance and partial pressure of carbon dioxide, *Plant Physiol.*, **62**, 670–4.
Wong, S. C., Cowan, I. R. and Farquhar, G. D. (1979). Stomatal conductance correlates with photosynthetic capacity., *Nature*, **282**, 424–6.

6 The control of stomata by hormones and other naturally occurring products

The effects of ABA on stomatal behaviour are well known. However, another group of hormones, the cytokinins, may also be involved in the control of stomata in some species. Other naturally occurring compounds including certain fatty acids, phenolics and simpler metabolites, such as proline, also affect stomata. Additionally phytotoxins such as fusicoccin and victorin, which are produced by fungal pathogens, affect stomata.

The effects of certain environmental factors on stomatal behaviour may be mediated by hormones. For example, water stress, salt stress and chilling of plants can result in elevated ABA levels within leaves with subsequent stomatal closure. Developmental changes in a plant can also indirectly influence stomatal behaviour probably by affecting hormonal levels within leaves. For example, the development of fruit in some species is paralleled with wider stomatal openings in the leaves and with changes in the levels of ABA, cytokinins or gibberellins within the leaves. Also, the previous growth conditions of a plant may influence hormone levels in leaves which, in turn, may affect stomatal behaviour. The 'after-effect' of wilting may be an example of this.

This chapter gives an account of what is known about the hormonal control of stomata at a physiological and a biochemical level and also discusses the effects of other naturally occurring products on stomatal behaviour. A section on the artificial control of stomata is also presented.

6.1 Abscisic acid

6.1a Stomatal responses to ABA

In the late 1960s it was discovered that ABA inhibited transpiration (Little and Eidt 1968; Mittelheuser and van Steveninck 1969) and

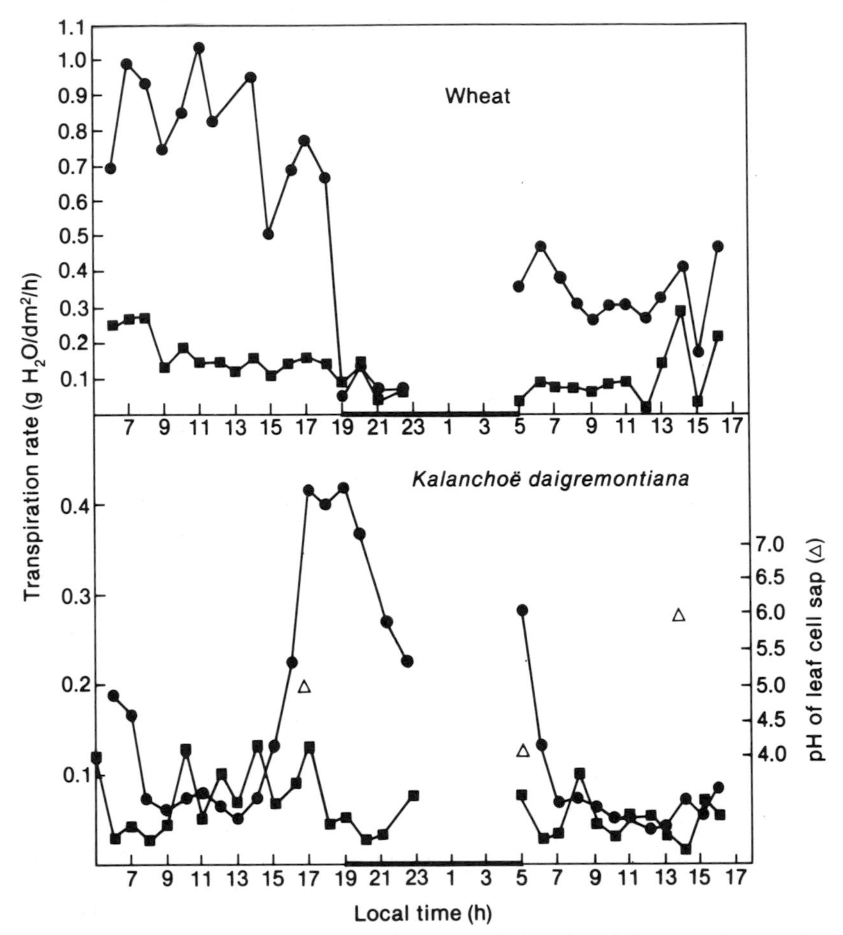

Figure 6.1 Transpiration rates of wheat and *Kalanchoe daigremontiana* with their petioles immersed in 10^{-4} M ABA (■) or water (●). The bar on the abscissa indicates the period of darkness (Scott-Murphy and Willmer, unpublished).

accumulated in leaf tissue under water stress (Wright 1969; Wright and Hiron 1969). It is now well established that ABA prevents stomatal opening and causes closure in virtually all plant species so far investigated. Night opening of stomata in *Kalanchoë daigremontiana*, a CAM plant, is also prevented by ABA (Fig. 6.1).

A few species of plants, however, are reported to be unresponsive to ABA. These species include yellow lupin (*Lupinus luteus*) and *Tradescantia virginiana*.

ABA will close stomata within minutes of application whether it is applied to the surface of leaves or via the transpiration stream of

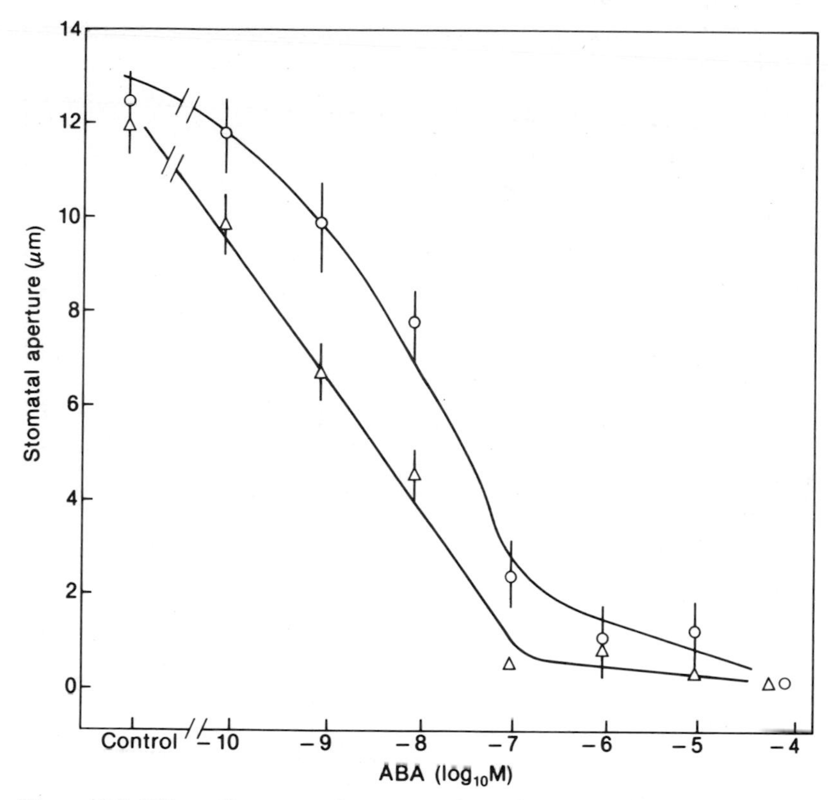

Figure 6.2 Effect of a range of concentrations of ABA in 50 mM tris-maleate buffer, pH 6.0, containing 300 mM KCl (○) and in 10 mM sodium citrate buffer containing 10 mM KCl, pH5.5 (△), on stomatal opening in epidermal strips of *Commelina communis*. Initially the stomata were closed and the epidermal strips were incubated in the light (500 W m^{-2}) at 30°C with air bubbling through the medium. Stomatal apertures were measured after 2 hours incubation. Vertical lines = std error of the means (where S.E. is smaller than the point vertical lines are absent) (from Willmer *et al.* 1978).

excised leaves; ABA will also prevent opening of stomata when supplied to leaves by either method. Additionally, ABA, under most circumstances, closes stomata and prevents opening in epidermal strips (Fig. 6.3). However, if the concentration of KCl in the incubation medium is increased the closing effects of ABA decrease. The major reasons for believing that ABA acts as a hormonal control of stomatal functioning are that exceedingly low concentrations affect stomatal behaviour (Fig. 6.2), that ABA produces its effects on stomata even in epidermal strips in the absence of the complexing influence of leaf mesophyll, that ABA causes guard cell protoplasts

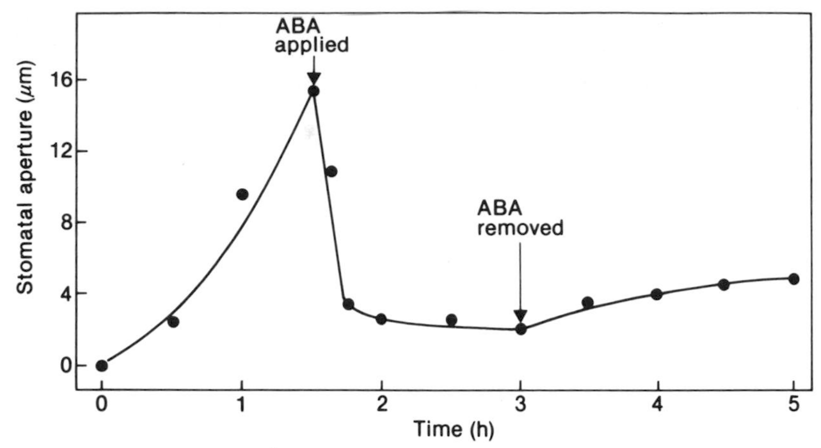

Figure 6.3 Effect of 10^{-4} M ABA on open stomata in epidermal strips of *C. communis* and their subsequent recovery in the absence of ABA. Strips were incubated in the light (500 Wm^{-2}) at 30°C with air bubbling through the medium (from Willmer *et al.* 1978).

to contract and that ABA levels build up in leaves as water stress occurs. Also, wilty mutants unable to synthesise ABA close their stomata upon an exogenous supply of ABA.

Some investigators, however, have doubted the involvement of ABA in the control of stomatal behaviour mainly because there is sometimes a poor correlation between bulk leaf water potential and ABA content of the leaves or between stomatal aperture and ABA content of the leaves. However, the possible reasons for the lack of correlation are manyfold. The amount of ABA synthesised by a plant may depend on the type of stress (whether osmotic, water or cold stress), the extent and duration of the stress, the previous growing conditions of the plant, the developmental stage of the plant (whether flowering or fruiting, for example) and on the inherent ability of a plant to manufacture and degrade ABA.

Under stress conditions, generally, ABA levels increase to the greatest extent in mesophytes and less so in xerophytes. Aquatic plants and particularly plants with submerged leaves and no stomata have a very limited ability to synthesise ABA under stress conditions. In some plants, such as *Tradescantia*, *Menyanthes* and *Mentha*, it has been reported that there was no increase in ABA levels of leaves under water stress.

In some species it has been observed that the bulk leaf water potential is not the best indicator of the water stress experienced by the epidermis and, therefore, the stomata of that leaf. Hence, there may not be a good correlation between bulk leaf ABA levels and stomatal aperture.

Another factor which may account for the lack of correlation between bulk leaf ABA levels and stomatal aperture or bulk leaf water potential is the possibility of a redistribution of ABA or the release and binding of ABA which may occur within a leaf during stress.

Perhaps even of more significance is the finding that leaf ABA levels appear more closely related to leaf *turgor* potentials rather than bulk leaf water potentials (Davies and Lakso 1978), and as turgor approaches zero ABA levels rise sharply (Pierce and Raschke 1980).

For these reasons it is difficult to obtain values for the threshold leaf water potentials that stimulate ABA synthesis. However, it has been concluded that only a few percent over the initial level of ABA is sufficient to initiate closure and that ABA can be synthesised rapidly enough to account for the observed rates of closure.

6.1b *After-effect of wilting*

If a water-stress period is relieved, in some species there is a period of days before stomata open to their full potential (Fig. 6.4). This is considered to be due to ABA slowly being disposed of (by degradation, or binding of ABA in the guard cells or slow exit from the guard

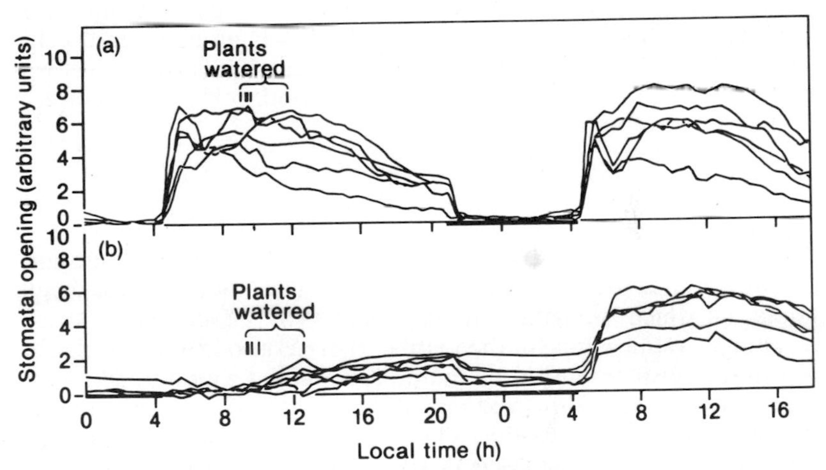

Figure 6.4 The after-effect of water stress on the stomatal behaviour in leaves of *Vicia faba*. (a) unstressed control plants; (b) plants were not watered for 3–6 days until the fourth expanded leaf from the apex had wilted for 2 days. At the points indicated by the short bars individual plants were watered. In each treatment the stomatal behaviour in six different leaves was recorded with a viscous flow porometer. Black bars on the abscissa indicate the dark periods (after Allaway and Mansfield 1970).

cells) thereby allowing only slow stomatal recovery. However, some authorities have disagreed with this explanation since they found that leaf ABA levels and stomatal recovery did not correlate well. For example, prestress ABA levels have been recorded in leaves before stomatal recovery and, in other cases, stomata have been found to open after application of ABA before appreciable break-down of ABA has occurred. Nevertheless, the lack of correlation between bulk leaf ABA levels and stomatal aperture may be due to one or more of the many reasons cited above.

6.1c CO_2/ABA interactions

Raschke (1975) suggested that there is an interdependence, at least in some species, between CO_2 and ABA in bringing about stomatal closure since he found that stomata of *Xanthium strumarium* did not close in the light in response to increased CO_2 concentrations unless the leaves had been fed ABA. Conversely, he found, stomata would not close if leaves were fed ABA unless CO_2 was present in the air. Mansfield (1976), however, using the same species, found no statistical interaction of ABA and CO_2 in their mode of action. However, the two reports do not necessarily conflict. The work of Raschke was carried out at a much higher light intensity than that of Mansfield and at high light intensities stomata in some species may become less responsive to stimuli. Undoubtedly ABA levels around guard cells and leaf CO_2 levels will play some interrelated role in controlling stomatal aperture since, individually, these both bring about closure as their concentrations increase.

6.1d Wilty mutants

Wilty tomato mutants (*flacca*, *sitiens* and *notabilis*), a pepper mutant (*Capsicum scabrous diminutive*) and a potato mutant have been found in which stomata remain open even under water stress resulting in rapid wilting of the plants. There is also an increased root resistance to water flow in the mutants compared with normal plants which contributes to the wilting of the plants. The reason for the stomata remaining open under plant water stress is that a single gene mutation results in decreased ABA synthesis. Secondary effects in the *flacca* mutant are the presence of high levels of auxin and kinetin-like activity which might stimulate stomatal opening. Exogenously supplied ABA brings about stomatal closure, thereby preventing wilting, although there is a report that short-term applications of ABA are ineffective at closing the stomata. Tal, Imber and their colleagues have been most active in research within this area (Imber and Tal 1970; Tal and Imber 1970).

6.1e Stomatal specificity to ABA

Generally, most investigators agree about the most active form of ABA and its related compounds which affect stomata. Difficulties are encountered, however, in establishing which structure is the most active since applied compounds may be metabolised to more, or less, active forms within the tissues and, depending on how polar a compound is, penetration of membranes may be slowed down or speeded up.

Stomata respond to the naturally occurring (+) enantiomer but very little to the unnatural (−) enantiomer. In some biological systems, however, both enantiomers are active. The C2-*cis*, C4-*trans* isomer is active while the C2-*trans*, C4-*trans* isomer is inactive. The presence of an O atom at the C-1 position appears essential for high activity although the complete 2,4-pentadience side chain does not appear essential for activity, while a ring carbonyl and hydroxyl group seem less essential for stomatal activity.

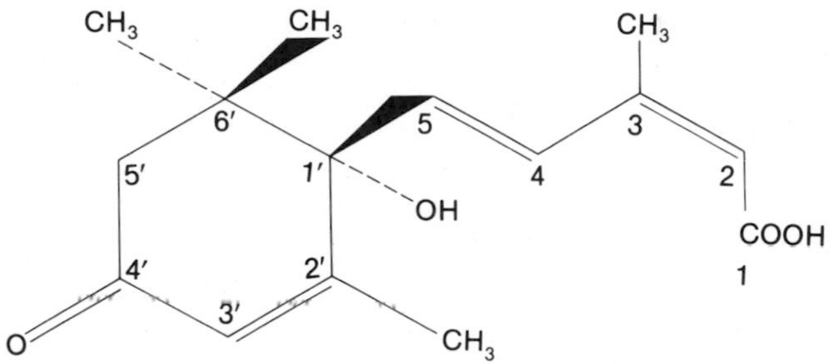

Figure 6.5 (+)-ABA. Note that C1′ is the only asymmetric C and this confers optical activity to the molecule. The (+) enantiomer rotates plane polarised light to the right while the (−) enantiomer rotates plane polarised light to the left.

6.1f Metabolism of ABA

ABA is a terpene (or terpenoid) compound. Terpenes are made up of multiples of five carbon isoprene units. The fifteen carbon compounds, of which ABA is one, are known as sesquiterpenoids. Figure 6.6 shows the two possible pathways of ABA synthesis and also the possible derivation of other sesquiterpenoids, such as farnesol, which affect stomata. There are two possible pathways of ABA biosynthesis known. The most popular view considers that ABA is formed from farnesyl pyrophosphate which in turn is ultimately derived from acetate. Carotenoids are also synthesised via farnesyl pyrophosphate and it is possible that ABA may be produced as a degradative product of certain carotenoids (particularly violaxanthin) via xanthoxin.

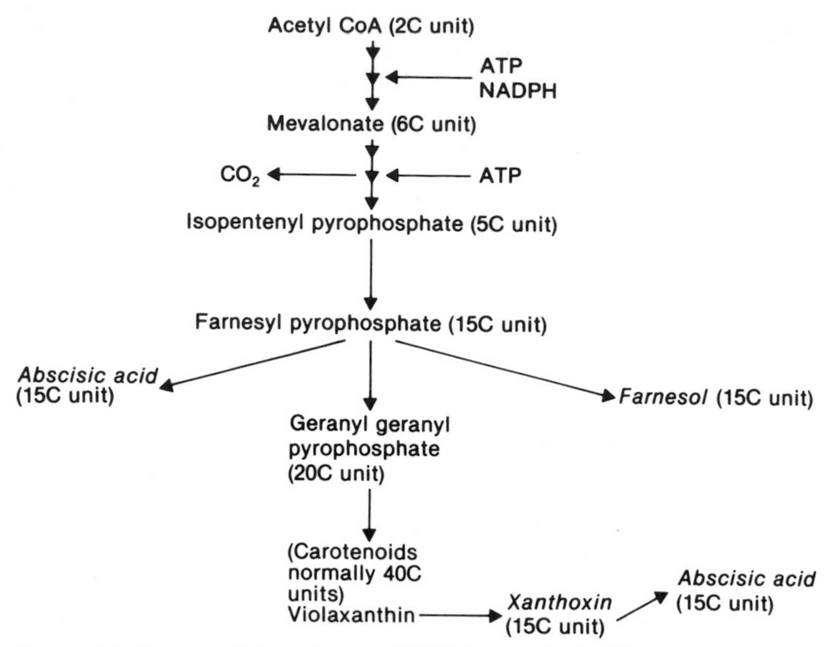

Figure 6.6 Two possible pathways of ABA biosynthesis. The compounds printed in italics have all been found to affect stomatal behaviour. Note that, except for mevalonate and acetyl CoA, all the compounds are multiples of five carbon units.

ABA is probably degraded mainly in the cytoplasm via 6′-hydroxylmethyl-ABA to phaseic acid and then to hydrophaseic acid. The α-D-glucose ester of ABA is also known to exist and may occur as a bound form of ABA within the cells.

6.1g Mode of action of ABA

Most reports indicate that ABA is synthesised within chloroplasts of the mesophyll tissue. It has generally been regarded that guard cells are unable to synthesise ABA although there is not complete agreement on this point. Loveys (1977) found that isolated mesophyll tissue of *Vicia faba* synthesised ABA when the tissue was water-stressed while epidermal tissue did not. Epidermes from previously stressed leaves, however, contained elevated ABA levels. This suggests that ABA is synthesised in the mesophyll chloroplasts and not within the guard cells or epidermal tissue as a whole. Based on this premise it is believed that, under stress conditions, ABA is released from the mesophyll chloroplasts, or other sites of bound ABA within the leaf, to eventually reach the guard cells. The route ABA takes to the guard cells may be apoplastic or symplastic. Since there are plasmodesmatal connections between mesophyll cells and neighbouring epidermal cells ABA could be transported symplastically from the mesophyll chloroplasts directly to the guard cell wall where, presumably, it would diffuse into the guard cell protoplast to elicit its activity. ABA could also move into the apoplast of the mesophyll cells and be swept along to the guard cells in the transpiration stream.

There is also evidence that there are a number of pools of ABA within a leaf: presumably ABA can exit from the chloroplast and be bound at various sites within the cells or even move to the epidermal layer where it may be 'stored' to be released when the need arises. The latter situation would allow for a more rapid stomatal response since the distance the ABA would travel to the target site would be shorter. Epidermal tissue of tulip and *Commelina communis* also has the ability to metabolise ABA and produce several related products.

The mechanism by which ABA is released from the chloroplast or from bound sites within other compartments of the cell is not understood. The concentration gradient across the chloroplast envelope will, to some extent, control diffusion rates out of the chloroplast. But perhaps of more importance will be the pH values of the various cell compartments. Membranes are highly permeable to the undissociated form of ABA (brought about by more acid pH values) but not to the anionic form (brought about by more alkaline pH values). Thus, the greater the pH difference between compartments the higher the accumulation of ABA in the more alkaline compartment which acts as an anion trap (Heilmann *et al.* 1980). On this basis the vacuole (pH $\simeq$ 5) is not likely to accumulate ABA whereas the cytoplasm(pH $\simeq$ 7) is more likely to although the stroma of the chloroplast in the light is about pH 8.0 and would, therefore, be expected to retain most of the ABA of a cell. However, the permeability of the chloroplast envelope may be modified by substances produced under water-stress. It has been suggested, for example, that farnesol (*see* §6.2) causes the chloroplast envelope to become 'leaky', allowing ABA to readily leave the chloroplast during water-stress conditions.

The mechanism by which ABA brings about stomatal closure once it reaches the guard cells is not fully understood. When stomata close, whether due to environmental factors or elevated levels of ABA within leaves, a number of guard cell activities are set in motion almost simultaneously, e.g. guard cell malate levels decrease, the levels of starch in guard cells increases, K^+ exits from the guard cells. It is difficult therefore to ascertain what the primary effect of ABA is on the stomatal mechanism.

There appears to be no single mode of action of ABA which brings about changes in plant processes. ABA is able to affect membrane-related processes in a relatively rapid fashion and also, over long term periods, to affect the expression of genes via transcriptional and/or non-transcriptional control. Membrane effects of ABA are difficult to interpret since it has been observed to stimulate K^+ uptake in some plant tissues and inhibit uptake in other tissues. Presumably the growth conditions of the plants, tissue pretreatment and the species and type of tissue used could all affect the ABA response and account for the different observations. However, generally, ABA inhibits K^+/H^+ exchange across membranes and it is possible that it influences such exchange across the plasmalemma (and/or perhaps the tonoplast) of guard cells during stomatal movements. ABA not only prevents stomatal opening but brings about stomatal closure so presumably, if it acts at the 'ion pump' level, it must not only inhibit guard cell K^+ import/H^+ export but also promote K^+ export/H^+ import during stomatal closure. A one-way valve system can be imagined in which ABA prevents K^+ from entering guard cells, thereby preventing stomatal opening, but it allows or promotes K^+ efflux from guard cells of open stomata to bring about closure. A report by MacRobbie (1980) that ABA does not markedly reduce influx of Rb^+ or Br^- (analogues of K^+ and Cl^-) but stimulates a transient efflux of the ions in *Commelina communis* epidermis (exposed to a pH of 3.9 for several hours to kill all subsidiary and epidermal cells) remains difficult to interpret.

There are conflicting reports about the characterisation of ATPases from epidermal tissue and how ABA affects their activity. For example, one report concludes that ABA stimulates membrane-bound Mg^{2+} activated ATPases from epidermal tissue of tobacco while another report concludes that ABA inhibits plasmalemma ATPase also from epidermal tissue of tobacco.

ABA has also been reported to increase H^+ influx in epidermal tissue of tobacco in light and dark and to increase malate efflux from guard cells during stomatal closure. Additionally, ABA has been observed to increase cell membrane permeability and to increase hydraulic conductivity of roots. If guard cell membranes are similarly affected then ABA could modify the speed of stomatal responses.

ABA could also control stomatal behaviour by acting at the metabolic level. ABA does not, however, affect phosphoenolpyrurate (PEP) carboxylase activity isolated from tulip leaf epidermal tissue. ABA does inhibit gibberellin-stimulated synthesis of α-amylase, thereby preventing starch hydrolysis in barley seeds. If starch hydrolysis in guard cells is inhibited in a similar fashion, then shortage of carbon skeletons for malate synthesis would result which, in turn, may prevent stomatal opening.

6.2 Other sesquiterpenoids (xanthoxin, all-*trans* farnesol, vomifoliol, phaseic acid)

Xanthoxin is about 50 per cent as active as ABA at causing stomatal closure when applied to leaves via the transpiration stream but inactive with epidermal strips of *Commelina communis* or *Vicia faba* (Raschke *et al.* 1975). Water stress does not increase xanthoxin levels within leaves. Xanthoxin may be a precursor of ABA (*see* Fig. 6.6) and the above observations would support this contention: presumably, when xanthoxin is applied to leaves via the transpiration stream, it is first converted to ABA in the mesophyll before it affects the stomata. There is no effect of xanthoxin on stomata in epidermal strips possibly because the epidermis cannot convert xanthoxin to ABA.

All-*trans* farnesol levels increase in water-stressed leaves of *Sorghum sudanense* (Wellburn *et al.* 1974). Farnesol also causes stomatal closure, which can be reversed, when applied to *Sorghum* leaves; it also causes closure and prevents opening of stomata in epidermal

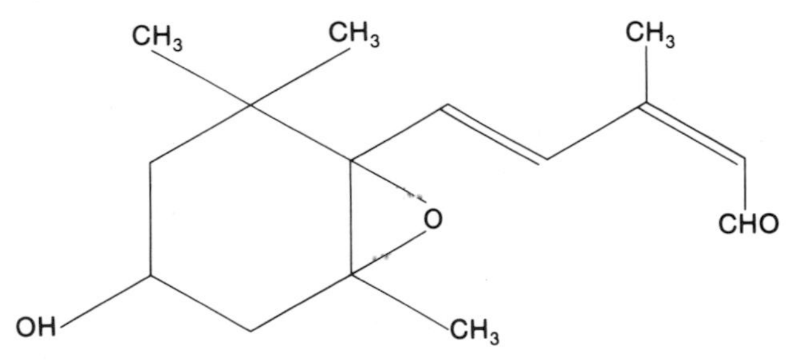

Figure 6.7 Xanthoxin.

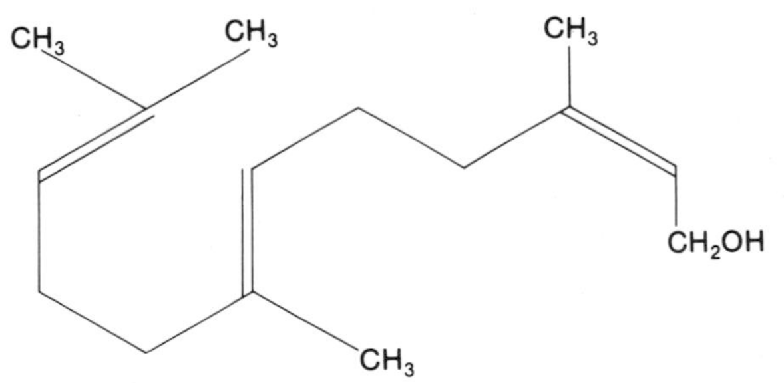

Figure 6.8 Farnesol.

strips of *Commelina communis*. Farnesol causes structural changes to cells damaging membranes, including chloroplast envelopes, at non-physiological concentrations (10^{-4}M and above).

Phaseic acid was considered at one time to be involved in the control of stomata since water stress and certain photoperiodic changes led to increased phaseic acid levels in leaves which were correlated with stomatal closure. The effectiveness of phaseic acid at closing stomata in epidermal strips and intact leaves varies from species to species. For example, it closes stomata in epidermal strips of *Commelina communis* but not of *Vicia faba*. One group of investigators found that the acid inhibited photosynthesis, thereby elevating leaf CO_2 levels which would close stomata, but other investigators found no inhibition of photosynthesis by phaseic acid. In general, the effects of phaseic acid on stomatal behaviour are not as marked as the effects of ABA and it is assumed that the former compound is a degradative product of the latter though retaining some activity. (+)-Vomifoliol is also reported to be as active as ABA on stomatal behaviour in epidermal strips of *Eichhornia* (Stuart and Coke 1975).

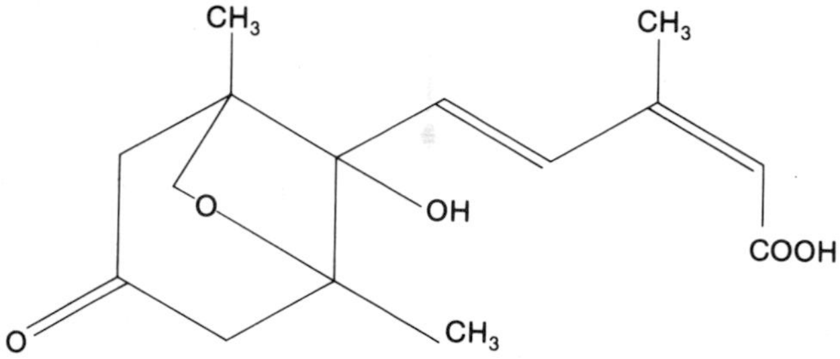

Figure 6.9 Phaseic acid.

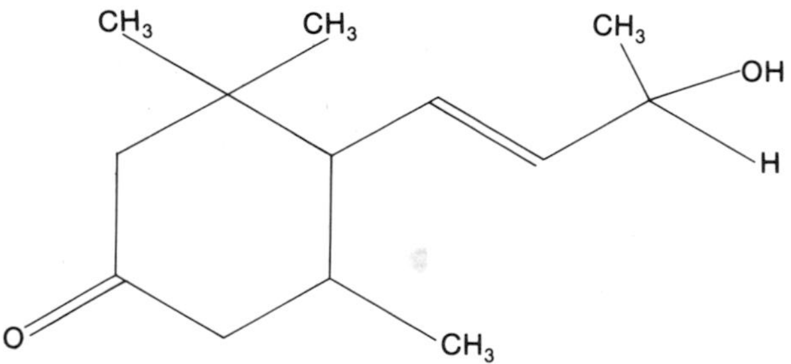

Figure 6.10 Vomifoliol.

6.3 Other hormones

6.3a Kinetin

Kinetin increases rates of transpiration in grass species but not other monocotyledons or dicotyledons. It has been suggested that the increased transpiration observed in grasses is due to kinetin delaying senescence of the excised leaves relative to rates of senescence in water controls. In turn, the delayed senescence will extend the CO_2 fixing ability of the leaves resulting in lower leaf CO_2 levels. As a consequence of the lowered CO_2 levels, stomata will open more widely in the kinetin treated leaves. Pallas and Box (1970), however, believe that the cytokinins act on stomata indirectly by affecting water potentials elsewhere in the plant.

Kinetin has no effect on stomata in epidermal strips of *Commelina communis* (*see* Table 6.1) but it does enhance stomatal opening in epidermal strips of the grass, *Anthephora pubescens* (Incoll and Whitelam 1977). Worthy of further investigation is a report that benzyl adenine reversed stomatal closure induced by ABA in epidermal strips of *Commelina benghalensis*.

It is apparent that the effects of cytokinins on stomata are not clear and little is known at present of their mode of action on stomata in those species it affects.

6.3b *Auxins*

Reports on the action of auxins are rather contradictory and difficult to evaluate. Some report that auxins can cause stomatal closure when applied to leaves or even roots. Others report no effects of auxins when applied to leaves. Zelitch (1961) found that when tobacco leaf discs were floated on solutions of the synthetic auxins, 2,4-D or NAA, stomata closed. However, auxin has no effect on stomata in epidermal strips of *Commelina communis* (*see* Table 6.1).

Although more studies are needed in this area, in general, the short-term effects of auxin on stomata do not seem very significant; the longer term effects may be more marked. It is likely, however, that if auxins do affect stomata it is in an indirect way by affecting the water relations or development of plants.

6.3c *Gibberellins*

Gibberellins have been reported to increase transpiration rates in excised barley leaves but most observers report no effect on stomata in intact leaves or epidermal strips (*see* Table 6.1).

6.3d *Ethylene*

CO_2 is a competitive inhibitor of ethylene action in many plant responses; the synthesis of ethylene has also been observed to increase in plants under water stress. Stomatal responses are also

Table 6.1 Effect of three growth regulators on stomatal opening in epidermal strips of *Commelina communis* when applied alone and together with abscisic acid. Mean apertures are based on 90–120 individual measurements (From Tucker and Mansfield 1971)

Treatments	Mean aperture (μm)	Standard error
Experiment 1		
Control	7.52	0.24
10^{-4} M ABA	0.73	0.08
10^{-4} M ABA + 10^{-3} M IAA	0.83	0.10
10^{-4} M ABA + 10^{-4} M IAA	0.62	0.07
10^{-4} M ABA + 10^{-5} M IAA	0.44	0.05
10^{-4} M ABA + 10^{-6} M IAA	0.48	0.06
10^{-5} M ABA	1.58	0.16
10^{-5} M ABA + 10^{-3} M IAA	1.27	0.13
10^{-5} M ABA + 10^{-4} M IAA	1.33	0.11
10^{-5} M ABA + 10^{-5} M IAA	1.33	0.15
10^{-5} M ABA + 10^{-6} M IAA	1.29	0.12
Experiment 2		
Control	6.84	0.24
10^{-3} M IAA	6.11	0.26
10^{-4} M IAA	6.92	0.26
10^{-5} M IAA	6.78	0.26
10^{-6} M IAA	6.78	0.28
Experiment 3		
Control	8.12	0.26
10^{-3} M Kinetin	8.28	0.22
10^{-4} M Kinetin	7.69	0.23
10^{-5} M Kinetin	7.48	0.26
10^{-6} M Kinetin	8.10	0.24
Experiment 4		
Control	7.86	0.02
10^{-4} M ABA	0.59	0.05
10^{-4} M ABA + 5×10^{-4} M Kinetin	1.00	0.09
10^{-4} M ABA + 5×10^{-5} M Kinetin	1.07	0.12
10^{-4} M ABA + 5×10^{-6} M Kinetin	0.59	0.07
10^{-5} M ABA	1.31	0.08
10^{-5} M ABA + 5×10^{-4} M Kinetin	1.22	0.09
10^{-5} M ABA + 5×10^{-5} M Kinetin	1.16	0.08

Table 6.1 *Continued*

Treatments	Mean aperture (μm)	Standard error
Experiment 5		
Control	7.12	0.26
10^{-4} M ABA	0.89	0.06
10^{-4} M ABA + 5 × 10^{-4} M GA_3	1.19	0.10
10^{-4} M ABA + 5 × 10^{-5} M GA_3	1.02	0.08
10^{-4} M ABA + 5 × 10^{-6} M GA_3	0.97	0.04
10^{-4} M ABA + 5 × 10^{-7} M GA_3	1.05	0.08
10^{-5} M ABA	1.39	0.11
10^{-5} M ABA + 5 × 10^{-4} M GA_3	1.58	0.03
10^{-5} M ABA + 5 × 10^{-5} M GA_3	1.42	0.08
10^{-5} M ABA + 5 × 10^{-6} M GA_3	1.18	0.06
10^{-5} M ABA + 5 × 10^{-7} M GA_3	1.23	0.07
Experiment 6		
Control	7.15	0.21
5 × 10^{-4} M GA_3	7.44	0.24
5 × 10^{-5} M GA_3	7.16	0.23
5 × 10^{-6} M GA_3	7.02	0.25
5 × 10^{-7} M GA_3	7.17	0.24

sensitive to changing CO_2 concentrations and water stress, and, hence, investigators have attempted to connect stomatal functioning and ethylene. However, Pallaghy and Raschke (1972) could find no effect of ethylene within the concentration range of 1 to 100,000 $\mu l\ l^{-1}$ on stomatal behaviour in corn or pea. Nor have ethylene effects been observed with stomata in sepals of carnations. Kays and Pallas (1980), however, found that in certain species, notably sunflower and peanut, exposure of the plants to 1.0 $\mu l\ l^{-1}$ ethylene for 2 hours resulted in substantial reduction of photosynthesis. Whether ethylene directly brought about stomatal closure thereby reducing photosynthesis or whether the ethylene effect on stomata was an indirect one due to the gas inhibiting photosynthesis which, in turn, would increase leaf CO_2 levels to close stomata, is not clear.

6.3e Phytochrome

Habermann (1973) concludes that stomatal movements of *Helianthus annuus* (wild type and *xantha* mutant) are phytochrome-mediated. However, not only were very low light intensities used in her experiments, but changes in stomatal aperture were only of the order of 1–2 μm. According to Evans and Allaway (1972) phytochrome is not involved in stomatal closure of *Vicia faba*.

6.4 Fatty acids

Short-chain saturated fatty acids (of carbon chain length C_6 to C_{11}) accumulate in leaves of barley and french bean under water stress. The C_9 (nonanoic), C_{10} (decanoic) and particularly C_{11} (undecanoic) acids also prevent stomatal opening and cause closure of stomata in epidermal strips of *Commelina communis*. The effectiveness of the short-chain fatty acids on stomatal behaviour is not as great as with ABA. There is a relatively narrow concentration range (between 10^{-6} and 10^{-4} M) over which they assert their action on stomata and stomatal closure is slow in their presence. Like farnesol, high concentrations (in the order of 10^{-4} M) cause cell and particularly membrane damage.

The longer chain unsaturated fatty acids, linolenic (C_{18}, 3 double bonds) and linoleic (C_{18}, 2 double bonds), also prevent stomatal opening.

Further investigation is needed into the effects of fatty acids to ascertain their significance in stomatal control.

6.5 Phytotoxins

A number of phytotoxins produced by plant pathogens affect stomata either by promoting or inhibiting opening. Little is known about most of the compounds but fusicoccin has been investigated in some depth.

Fusicoccin is a phytotoxin produced by submerged cultures of the fungal pathogen, *Fusicoccum amygdali*. The substance opens stomata when applied to leaves or epidermal strips and can overcome the closing effects of ABA (Squire and Mansfield 1972).

Fusicoccin increases H^+ extrusion and K^+ uptake in many types of cells and tissues and also hyperpolarises membrane potentials. Fusicoccin may, therefore, bring about stomatal opening by influencing ion transport across the plasmalemma of guard cells.

Victorin (a pentapeptide joined to a secondary amine) is a phytotoxin produced by the fungus *Helminthosporium victoriae* which causes stomatal closure. Little more is known about its effects on stomata.

A toxin produced by *Helminthosporium maydis*, race T, which causes southern corn leaf blight also causes solute leakage from roots of susceptible plants and induces rapid stomatal closure possibly by inhibiting K^+ uptake (Arntzen *et al.* 1973).

6.6 Phytoalexins

Phytoalexins are antimicrobial substances produced by certain host plants normally in response to pathogen attack. Many of these

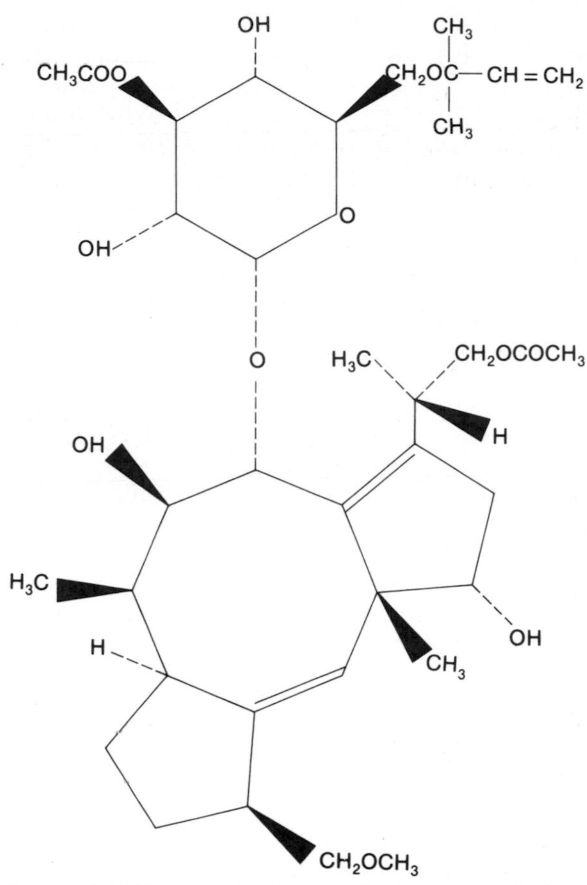

Figure 6.11 Fusicocccin I, a complex molecule which can overcome the stomatal closing effects of ABA.

compounds have been found to affect stomatal behaviour. For example, wyerone acid, a phytoalexin produced by *Vicia faba*, inhibits stomatal opening in *V. faba* and *Commelina communis* epidermal strips thoughh not as effectively as ABA. Other phytoalexins, such as pisatin and phaseollin, have similar effects on stomata in epidermal strips of *V. faba* and *Commelina communis* (Plumbe and Willmer, unpublished). Pisatin, a phytoalexin produced by peas, has been observed to inhibit stomatal opening in pea leaves due to the inability of the guard cells to accumulate solutes (Ayres 1980).

Many phytoalexins damage membranes. Their effects on stomata may also be due, mainly, to changing membrane permeability of guard cells.

Investigations so far indicate that phytoalexins are not produced by plants grown under water stress in the absence of pathogens unless, in some cases, tissue death occurs. However, certain compounds produced under water stress such as short chain fatty acids and farnesol, inhibit spore germination of *Cladosporium herbarum*; ABA has no effect on the germination of spores of *C. herbarum* or the growth of their germ tubes (Plumbe and Willmer, unpublished).

6.7 Other compounds

A variety of other naturally-occurring compounds of a diverse nature have been found to affect stomata. Usually studies have not been detailed enough to determine the importance or significance of the compound in the control of stomatal behaviour. Also, usually the effects on stomata have not been great and relatively high concentrations are needed to elicit a stomatal response.

A number of phenolic compounds affect stomata. Ferulic acid, for example, was found to increase transpiration rates in barley leaves but it has little effect on stomatal opening in *Commelina communis* epidermal strips. Sinapic acid and caffeic acid inhibit stomatal opening in epidermal strips of *Commelina communis* and reduce transpiration rates in excised leaves.

The amino acid, proline, shows some activity at inhibiting stomatal opening in epidermal strips of *Commelina communis*. Proline also accumulates in large amounts in leaves of some species exposed to water stress. This correlation between water stress, proline accumulation and inhibition of stomatal opening deserves further attention.

6.8 Artificial control of stomata

The artificial control of stomatal behaviour could be of use to man in a number of ways. Antitranspirants could be used over the long term or short term to close stomata, enabling a crop to withstand a period of drought. Such treatment would be particularly useful with perennial crops where survival may be more important than good yield for a particular season (since closing the stomata would prevent photosynthesis, and reduce yield as well as preventing water loss). Antitranspirants may also be sprayed on newly transplanted trees and other plants during dry periods to cut down water loss from the leaves and increase their chances of survival. It is also enticing to think that it may one day be possible to extend the opening period of stomata, possibly in late afternoon, under ideal photosynthesising conditions when water was readily available. The daily period of photosynthesis would be extended and crop yields might, therefore, increase. For this sort of stomatal control protranspirants would be needed rather than antitranspirants and they would possibly have to overcome stomatal closing in late afternoon due to endogenous rhythms. Fusicoccin, for example, may be used as a protranspirant and it has already been used at an experimental level to speed up the drying of newly cut grass to be used for hay. Compounds may also be applied to plants to close stomata and prevent entry of toxic pollutants which may damage tissues within the leaf.

Such compounds which will open or close stomata must be non-toxic to man and the plant, cheap to produce and must act specifically on guard cells without directly affecting photosynthesis or other plant processes in the underlying mesophyll. A danger of using antitranspirants may be one of overloading the cooling capacity of leaves since evaporation and escape of water vapour from the leaves will be reduced.

To date there is no substance (synthetic or natural) which has a higher permeability to CO_2 than to water vapour (Raschke 1979). Also, because the water vapour gradient is steeper than the CO_2 gradient across the stomata more water molecules will be lost from the leaf than CO_2 will enter the leaf. Thus, no antitranspirant can reduce water loss without also reducing CO_2 uptake by the leaf. However, a reduction in stomatal conductance affects transpiration and assimilation to different degrees and the ratio between the rates of the two processes (transpiration/assimilation, the transpiration ratio) usually declines as stomata close.

There are basically two types of antitranspirants. One type may be classed as film antitranspirants. These are often silicone-based oils or emulsions of latex or waxy substances which are sprayed on crops to give a semi-waterproof coating. The other type of antitranspirants are sprays of either naturally occurring or synthetic compounds which directly affect stomata.

Many compounds have been considered as possible antitranspirants but most have deleterious side effects. Phenylmercuric acetate has been used in the past with some success, even in field trials, but, of course, it is a serious environmental pollutant. Some of the possible natural occurring antitranspirants unfortunately are likely to have side effects. ABA, however, is needed at very low concentrations to close stomata and, when sprayed on plants, appears to have minimal affects on other plant processes. The disadvantages of using ABA as an antitranspirant may be its short-lived effect and possibly its expense. Recently it has been found that acetylsalicylic acid (aspirin) prevents opening in epidermal strips. This observation may lend some support to the belief that the addition of aspirin to water containing flowers preserves the flowers longer than in the absence of aspirin. Although this needs further verification in whole plants, presumably the aspirin will be taken up in the transpiration stream and close the stomata in the foliage (and possibly petals and other floral parts), thereby preventing excessive water loss and wilting. If this is the case, aspirin deserves further attention as a possible antitranspirant.

Although there are a few impressive reports on the effects of film antitranspirants, results generally have not been successful. Film antitranspirants are difficult to use with precision and may even cause leaf necrosis.

The search for ideal antitranspirants continues. New, naturally occurring and synthetic compounds which affect stomata are still being discovered. Some of these may prove to be useful for artificially controlling stomatal behaviour in the field.

Further reading and references

Allaway, W. G. and Mansfield, T. A. (1970) Experiments and observations on the after-effect of wilting on stomata of *Rumex sanguineus*, *Can. J. Bot.*, **48**, 513–21.

Arntzen, C. J., Haugh, M. F. and Bobick, S. (1973) Induction of stomatal closure by *Helminthosporium maydis* pathotoxin, *Plant Physiol.*, **52**, 569–74.

Ayres, P. G., (1980) Stomatal behaviour in mildewed pea leaves: solute potentials of the epidermis and effects of pisatin, *Physiol. Plant Pathol.*, **17**, 157–65.

Davies, F. S. and Lakso, A. N. (1978) Water relations in apple seedlings: changes in water potential components, abscisic acid levels and stomatal conductance under irrigated and non-irrigated conditions, *J. Am. Soc. Hort. Sci.*, **103**, 310–13.

Evans, L. T. and Allaway, W. G. (1972) Action spectrum for the opening of *Albizzia julibrissin* pinnules and the role of phytochrome in the closing movements of pinnules and of stomata of *Vicia faba*, *Aust. J. Biol. Sci.*, **25**, 885–93.

Habermann, H. M. (1973) Evidence for two photoreactions and possible involvement of phytochrome in light-dependent stomatal opening, *Plant Physiol.*, **51**, 543–8.

Heilmann, B., Hartung, W. and Gimmler, H. (1980) The distribution of abscisic acid between chloroplasts and cytoplasm of leaf cells and the permeability of the chloroplast envelope for abscisic acid, *Z. Pflanzenphysiol*, **97**, 67–8.

Imber, D. and Tal, M. (1970) Phenotypic reversion of *flacca*, a wilty mutant of tomato, by abscisic acid, *Science*, **169**, 592–3.

Incoll, L. O. and Whitelam, G. C. (1977) The effect of kinetin on stomata of the grass *Anthephora pubescens* Nees, *Planta*, **137**, 243–5.

Kays, S. J. and Pallas, J. E. (1980) Inhibition of photosynthesis by ethylene, *Nature*, **285**, 51–2.

Little, C. H. A. and Eidt, D. C. (1968) Effect of absicic acid on bud break and transpiration in woody species, *Nature*, *220*, 498–9.

Loveys, B. R. (1977) The intracellular location of abscisic acid in stressed and non-stressed leaf tissue, *Physiol. Plant.*, **40**, 6–10.

MacRobbie, E. (1980) Stomatal ionic relations, in Spanswick, R. M., Lucas, W. J. and Dainty, J. (eds), *Plant Membrane Transport: Current Conceptual Issues*. Elsevier, Holland.

Mansfield, T. A. (1976) Delay in the response of stomata to abscisic acid in CO_2-free air, *J. Exp. Bot.*, **27**, 559–64.

Marrè, E. (1979) Fusicoccin: a tool in plant physiology, *Ann. Rev. Plant Physiol.*, **30**, 273–88.

Mittelheuser, C. J. and Van Steveninck, R. F. M. (1969) Stomatal closure and inhibition of transpiration induced by (RS)-abscisic acid, *Nature*, **221**, 281–2.

Pallaghy, C. K. and Raschke, K. (1972) No stomatal response to ethylene, *Plant Physiol.*, **49**, 275.

Pallas, J. E. and Box, I. E. (1970) Explanation for the stomatal response of excised leaves to kinetin, *Nature*, **227**, 87–8.

Pierce, M. and Raschke, K. (1980) Correlation between loss of turgor and accumulation of abscisic acid in detached leaves, *Planta*, **148**, 174–82.

Raschke, K. (1975) Simultaneous requirement of carbon dioxide and abscisic acid for stomatal closing in *Xanthium strumarium* L., *Planta*, **125**, 243–59.

Raschke, K. (1979) Movements of stomata, in Hampt, W. and Feinlieb, M. E. (eds), Physiology of Movements, *Encyclopedia of Plant Physiology*, New Series, vol. 7. Springer-Verlag, Berlin.

Raschke, K., Firn, R. D. and Pierce, M. (1975) Stomatal closure in response to xanthoxin and abscisic acid, *Planta*, **125**, 149–60.

Squire, G. R., and Mansfield, T. A. (1972) Studies of the mechanism of action of fusicoccin, the fungal toxin that induces wilting, and its interaction with abscisic acid, *Planta*, **105**, 71–8.

Stuart, K. L. and Coke, L. B. (1975) The effect of vomifoliol on stomatal aperture, *Planta*, **122**, 307–10.

Tal, M. and Imber, D. (1970) Abnormal stomatal behaviour and hormonal imbalance in *¾acca*, a wilty mutant of tomato. II. Auxin and abscisic acid-like activity. *Plant Physiol.*, **46**, 373–76.

Tucker, D. J. and Mansfield, T. A. (1971). A simple bioassay for detecting 'antitranspirant' activity of naturally occurring compounds such as abscisic acid, *Planta*, **98**, 157–63.

Walton, D. C. (1980) Biochemistry and physiology of abscisic acid, *Ann. Rev. Plant Physiol.*, **31**, 453–89.

Wellburn, A. R., Ogunkanmi, A. B., Fenton, R. and Mansfield, T. A. (1974) All-*trans*-farnesol: a naturally occurring antitranspirant?, *Planta*, **120**, 255–63.

Willmer, C. M., Don, R. and Parker, W. (1978) Levels of short-chain fatty acids and of abscisic acid in water-stressed and non-stressed leaves and their effects on stomata in epidermal strips and intact leaves, *Planta*, **139**, 281–7.

Wright, S. T. C. (1969) An increase in the 'inhibitor-β' content of detached wheatleaves following a period of wilting, *Planta*, **86**, 10–20.

Wright, S. T. C. and Hiron, R. W. P. (1969) (+) Abscisic acid, the growth inhibitor induced in wheat leaves by a period of wilting, *Nature*, **224**, 719–20.

Zelitch, I. (1961) Biochemical control of stomatal opening in leaves, *Proc. Natl. Acad. Sci.*, U.S.A., **47**, 1423–33.

7 The mechanism of stomatal functioning

Many ingenious hypotheses have been forwarded over the last century and a half to explain how stomata open and close. All have proved inadequate to lesser or greater extents. However, since the late 1960s major advances in our understanding of how stomata function have been made and our knowledge about the biochemistry of guard cells and of ion fluxes which occur between guard cells and neighbouring cells has increased enormously.

This chapter first outlines the various hypotheses that have been proposed to explain stomatal movements and points out their weaknesses (for further reading and references consult Heath and Mansfield 1969). The more modern views up to current thinking on the stomatal mechanism will then be presented. This will involve discussions on the biochemistry of guard cells, the role of ion fluxes in stomatal movement, guard cell energy sources and the electrophysiology of guard cells. A simplified summary of the complex events occurring in guard cells during stomatal movements is also presented at the end of the chapter.

7.1 Early hypotheses

7.1a The photosynthetic hypothesis

One of the earliest hypotheses to explain how stomata functioned was proposed by von Mohl (1856). He suggested that the products of guard cell photosynthesis were responsible for the observed osmotic potential changes in guard cells. However, stomata can open in the absence of atmospheric CO_2 and in the dark when photosynthesis cannot occur. Also, in some species, chloroplasts are not present in guard cells (*see* §3.3a) and in other species, although chloroplasts are present, the Calvin cycle may not operate (*see* §7.3b).

7.1b The starch ⇌ sugar hypothesis

The starch ⇌ sugar hypothesis has probably received most attention and support over the years to explain stomatal functioning. The idea was first conceived by Kohl (1895) and subsequently developed by Lloyd (1908). Later, others enlarged and embellished the starch ⇌ sugar hypothesis (see Heath and Mansfield 1969).

Essentially, the hypothesis was considered to work as follows: In the light, when stomata are normally open, photosynthesis reduces the leaf CO_2 levels which was considered to increase pH values within guard cells. Since high pH values favour the conversion of starch to sugar in a reaction catalysed by phosphorylase it was considered that sugars accumulated in the guard cells, resulting in a decreased osmotic potential. In the dark, when stomata normally close, the reverse sequence of events was considered to occur: carbon dioxide levels within the leaf and guard cells will increase, pH values within the cells will fall, and, under these conditions, phosphorylase favours the conversion of sugars to starch which is virtually osmotically inactive. There is considerable circumstantial evidence to support the hypothesis: pH values of guard cells decrease when stomata close and increase when stomata open (*see* Table 7.3); starch levels in guard cells generally decrease upon stomatal opening and increase upon closure; and phosphorylase activity has been detected in guard cells. However, there is also considerable evidence which does not support the hypothesis. First, although our knowledge of plant phosphorylase is fragmentary, pH may not be the deciding factor in determining whether the enzyme degrades or synthesises starch. The enzyme is located in plastids and is capable of forming or breaking α-1,4 bonds in both amylose and amylopectin in the reversible reaction:

$$\underset{\text{glucose-1-phosphate}}{\text{G-1-P}} + \alpha\text{-glucan 'primer'} \longleftrightarrow \underset{\text{(starch)}}{\alpha\text{-glucan}} + \underset{\text{inorganic phosphate}}{P_i}$$

The free energy change of the reaction indicates that it will go to the right to synthesise starch, particularly as pH values become more acidic. Even at pH 8.0, which is the value of the stroma of illuminated chloroplasts (and the location of phosphorylase), the forward reaction is energetically more favourable. However, the reaction is also dependent on the P_i:G-1-P ratio and, although the P_i levels in illuminated and darkened chloroplasts vary, there is probably always more P_i than G-1-P present. This will push the reaction to the left and favour starch break-down. Thus, the overall direction of the reaction will depend on the pH values *and* the P_i:G-1-P ratios of the stroma of guard cell chloroplasts when stomata are opening and closing.

Even in the direction of starch break-down the reaction cannot, of itself, lead to a decrease in guard cell osmotic potential since one osmotically active particle, P_i, has been used to create another, namely glucose-1-phosphate (the primer molecule and starch are essentially insoluble and, therefore, osmotically inactive). To obtain a net decrease in osmotic potential during stomatal opening G-1-P must be hydrolysed to liberate P_i and glucose. Upon stomatal closure the glucose must be phosphorylated before phosphorylase can convert it to starch.

Other criticisms have been aimed at the hypothesis. It has been suggested that the relatively large pH changes that have been observed to occur in guard cells are too large to be accounted for by changes of CO_2 concentrations experienced by the cells. It must be appreciated, however, that the observed pH changes which occur in guard cells are probably vacuolar values and not chloroplastic or cytoplasmic ones and, as already indicated, phosphorylase is located in the chloroplasts.

It has also been suggested that the changes in guard cell starch levels which occur during opening and closing movements are too slow to account for the relatively rapid stomatal movements although this was not demonstrated to be the case by all investigators. Also, starch changes in guard cells and stomatal aperture changes do not always correspond in the expected manner (*see* §7.3).

The starch ⇀ sugar hypothesis could not apply to the stomatal mechanism in those species which lack guard cell starch. However, in these species, fructans are possibly present in guard cells which could serve the purpose which starch (glucans) do in other species (*see* §7.3b).

Thus the starch ⇀ sugar hypothesis, as it exists, is too simplistic to be totally acceptable although the basic concept of the starch/sugar interconversion is not totally wrong; current thinking suggests that carbon flow occurring during stomatal movements is between starch, sugar and malate (*see* §7.3).

7.1c The permeability hypothesis

In the early 1900s a permeability hypothesis was proposed to explain stomatal movements. Guard cell membrane permeability was gauged by the rate of dye uptake by cells in the light or dark. Although there was some conflict of view as to whether guard cells were more permeable in the light or dark it was generally accepted that uptake was greatest in the light. Thus, it was concluded that solutes entered guard cells more readily in the light to decrease their osmotic potential and bring about stomatal opening. Unfortunately, scientists were unaware of the phenomenon of active transport at that time and under passive forces alone changing membrane permeabilities would only speed up or slow down equilibrium solute concentrations. Moreover, this hypothesis does not take into consideration stomatal closure in the dark when guard cell membrane permeability is decreased; if solutes are prevented from escaping from the guard cells their osmotic potential will not increase to enable stomata to close. Nevertheless, in view of what we now know about K^+ and other ion fluxes across the guard cell plasmalemma and tonoplast there is some semblance of validity about the hypothesis. Certainly the changes in membrane permeability due to light or dark could activate or modify a 'pumping' system in the guard cell plasmalemma and/or tonoplast.

7.1d The glycollate hypothesis

A scheme to explain stomatal movements which was prevalent in the 1960s was the so-called glycollate hypothesis proposed by Zelitch (1963). It stemmed from his observations that glycollate stimulated stomatal opening in tobacco leaf discs and that α-hydroxysulphonates (which inhibit glycollate oxidase, among other things) inhibited this opening. It was suggested that a glycollate $\rightleftharpoons$ glyoxyllate system was linked to non-cyclic photophosphorylation. ATP generated in the system could then be used to drive active ion 'pumps' integral to the stomatal mechanism. Alternatively, glycollate could be used as a source of carbon for production of carbohydrates available to decrease guard cell osmotic potentials.

The glycollate hypothesis, however, suffers from the criticism aimed at the photosynthetic hypothesis, namely that stomata can open in the dark in the absence of CO_2 when the light reactions and the CO_2 fixation processes of photosynthesis cannot occur. Furthermore, the guard cells of some species do not possess chloroplasts, the major site of glycollate production, and peroxisomes, the site of glycollate metabolism in the photorespiratory pathway, have never been unequivocally identified in guard cells. Also, the α-hydroxysulphonate inhibitors used by Zelitch are not specific inhibitors of glycollate oxidase. They are known, for example, to inhibit numerous steps in photosynthesis and so CO_2 levels would increase in the leaf tissues, bringing about stomatal closure.

7.1e The amphoteric colloid hypothesis

Another hypothesis of historical interest is one proposed by Scarth (1929), which may be called the amphoteric colloid hypothesis. Scarth proposed that in the light the increase in guard cell pH which occurs when stomata open results in the swelling of amphoteric colloids within the cytoplasm and increased guard cell turgor. In the dark the reverse situation was considered to occur with a contraction of the colloids, a decrease in guard cell turgor and stomata closed. However, although the cytoplasm of guard cells appears to change in viscosity and granularity as stomata open and close there is no direct evidence to support this hypothesis. Also, the pH of the guard cell cytoplasm is likely to remain fairly constant during stomatal movements while the vacuolar pH of the cells changes (*see* §7.2d).

All of the above hypotheses have now been discarded in part or in totality and a mechanism to explain stomatal movements based on the fluxes of K^+ and H^+ across guard cells and upon organic acid synthesis in guard cells is now the centre of attraction.

7.2 Ionic relations of stomatal movements

7.2a Potassium

In 1905, Macallum localised K^+ in guard cells of tulip with a histochemical test using sodium cobaltinitrite. It was not stated whether stomata were open or closed and the significance of this finding was lost until the work of Japanese scientists was published nearly half a century later. Imamura (1943) and Yamashita (1952) detected accumulation of K^+ in guard cells of a variety of species when stomata opened. However, it was not until Fujino (1967) published his work that scientists outside Japan became fully aware of the involvement of K^+ in stomatal functioning.

Fischer (1968) and his colleagues independently confirmed the findings of the Japanese workers using epidermis of *Vicia faba* and concluded that enough K^+, together with an equivalent amount of anion, could account for the decreased osmotic potentials of guard cells which occur when stomata open. They also proposed that the guard cell uptake mechanism was specific for K^+ and that it was light-activated.

About the same time Willmer and Mansfield (1969) found that a large number of monovalent alkali cations could stimulate stomatal opening in *Commelina communis* epidermis. However, the high concentration of cation used by Willmer and Mansfield in their studies was thought to cause lack of sensitivity to the ion species and result in non-physiological responses. Nevertheless, under saline conditions at least one halophyte, *Cakile maritima*, is known to use Na^+ rather than K^+ as the major osmoticum accumulated in guard cells (Eshel *et al.* 1974) It would be interesting, therefore, to determine if other halophytes utilise Na^+ in a similiar manner. Potassium is normally the most abundant inorganic element in leaves and therefore, not surprisingly, it is almost universally used to assist in decreasing and increasing guard cell osmotic potentials regardless of the evolutionary level of plants or the location of stomata on plants. For example, K^+ has been observed to accumulate in guard cells upon stomatal opening in leaves of *Equisetum* (horsetails), ferns, herbaceous plants and woody (including *Gingko biloba*) and non-woody perennials. When stomata open K^+ also accumulates in guard cells of stomata located in awns, sepals, and stems. Additionally, K^+ accumulates in guard cells of CAM plants during night opening. This suggests that lowered CO_2 levels as well as light can stimulate K^+ accumulation by guard cells since the dark opening of stomata in CAM plants is considered to be due to lowered CO_2 levels within the leaves (*see* §5.3a). Contrary to the findings of Nelson and Mayo (1979), K^+ has been observed to accumulate in guard cells of *Paphiopedilum* (lady slipper orchids) species and a variegated cultivar of *Pelargonium zonale* both of which do not possess guard cell chloroplasts but have functional stomata.

Most of the data on K^+ fluxes initially came from the sodium cobaltinitrite histochemical test, but later more quantitative data were obtained using electron microprobe X-ray analysis (Sawhney and Zelitch 1969; Humble and Raschke 1971; Raschke and Fellows 1971), K^+ sensitive microelectrodes (Penny and Bowling 1974), flame photometry coupled with 'rolling' of the epidermis which preferentially bursts the epidermal cells (Allaway and Hsiao 1973) and by a K^+ enhanced enzyme assay of extracts from single guard cell pairs (Outlaw and Lowry 1977). Table 7.1 shows the value for K^+ concentrations in guard cells of a number of species at various stomatal apertures. In general, values of around 500 mM K^+ are found in guard cells for open stomata (although this value will depend on the stomatal aperture) while values of closed or nearly closed stomata are usually less than 200 mM. The rates of the K^+ fluxes into guard cells are very high and for *Zea mays* Raschke (1975) observed a value of 10^{-14} eq min^{-1} which he determines is about twelve times the maximum velocity of the system 2 mechanism of ion uptake by barley roots.

The K^+ fluxes involved in the stomatal mechanism are probably restricted to the cells of the epidermal layer. The results from a simple experiment suggest this is the case for *Commelina communis*: when epidermis is floated on water and the tissue exposed to light and CO_2-free air opening does not occur; however, when similar pieces of epidermis are suspended in a humid atmosphere and exposed to the same conditions stomata open and K^+ enters the guard cells. Furthermore, in a number of species K^+ levels in the epidermis are almost twice that of the mesophyll. This information indicates that there is enough K^+ stored in the epidermis to support stomatal opening.

When stomata are closed, in species other than the grasses, K^+ is dispersed throughout the epidermal layer though often it is concentrated in cells surrounding the guard cells (*see* Figs 7.1, 7.2). Much of the K^+ is within the protoplasts though some may remain in the cell walls. Upon stomatal opening K^+ accumulates in the guard cells and

Table 7.1 Concentrations of K^+ in guard cells of open and closed stomata

Species	K^+ concentration (mM) in a guard cell of a stoma		Method of measurement	Reference source
	closed	**open**		
Vicia faba	77(2 μm)	883(12 μm)	microprobe	Humble and Raschke (1971)
	110*	554	'rolling' of epidermis and flame photometry	Allaway and Hsiao (1973)
	144*	536	'rolling' of epidermis and flame photometry	Allaway (1973)
	219†	614(10 μm)	K^+ enhanced enzyme assay	Outlaw and Lowry (1977)
Nicotiana tabacum	210(1 μm)	500(8 μm)	microprobe	Sawhney and Zelitch (1969)
Commelina communis	95	448	K^+ sensitive microelectrodes	Penny and Bowling (1974)
Zea mays	150	400	microprobe	Raschke and Fellows (1971)
Tradescantia albiflora	152	633	K^+ sensitive microelectrodes	Zlotnikova *et al.* (1977)

* No allowance made for changes in guard cell volume when stomata are open and closed. Guard cell volume of 5 pl assumed.
† Guard cell volume taken as 4.2 pl when stomata open and 2.2 pl when stomata close.

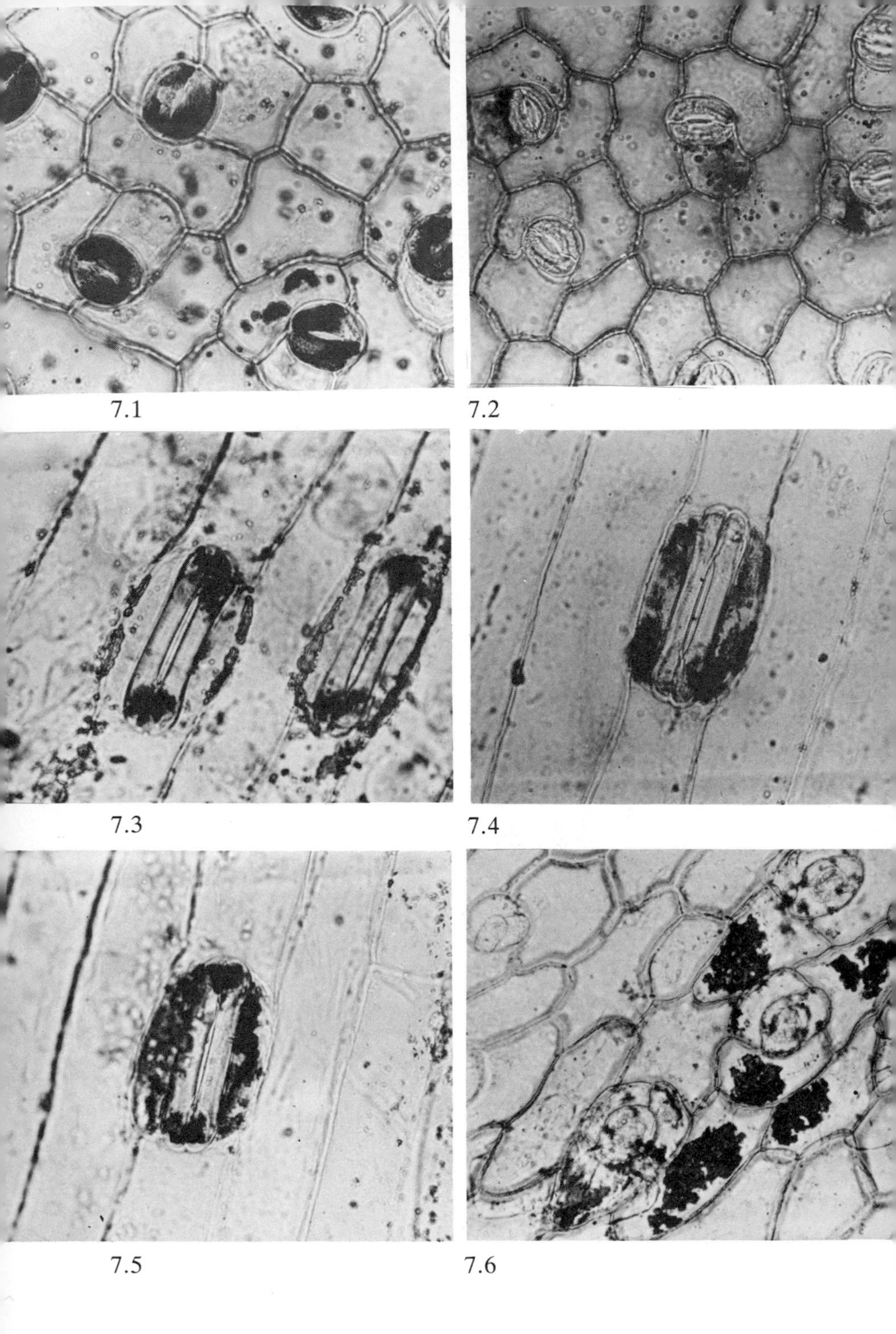

7.1

7.2

7.3

7.4

7.5

7.6

much of the K^+ may be transported across a number of cells to reach the guard cells, as in *Commelina* species. Although K^+ would be expected to accumulate in the vacuole of cells there are occasional reports of considerable amounts in the cytoplasm. In the grasses there is a shuttle of K^+ between subsidiary cells and guard cells; during opening K^+ accumulates in the guard cells while during the closure K^+ moves to the subsidiary cells (*see* Figs 7.3–7.5).

Occasionally epidermal cells located distantly from the guard cells accumulate high K^+ levels (Fig. 7.6). Although involved in storage of ions the significance of this is not understood.

In certain ferns and in *Tradescantia pallidus* and *T. virginiana* extracellular regions at the polar ends of guard cells which appear to accumulate K^+ have been reported (Stevens and Martin 1977). Their significance in the movement of K between guard cells and neigbouring cells is also not understood although they could be involved in ion storage.

7.2b Other inorganic cations

Monovalent cations other than K^+ have been observed to affect stomata. For example, Imamura (1943) found that, at low concentrations, cations stimulated stomatal opening in the following order of activity, $K > Na > Li$, while at higher concentrations the order was reversed. Iljin (1957) also observed that monovalent cations stimulated opening in the following order of activity, $Li > Na > Cs > K > Rb$, while Willmer and Mansfield (1969) found that many alkali monovalent cations stimulated stomatal opening in *C. communis* epidermis and that Na^+ and Li^+ were most effective. However, as indicated in §7.2a it was considered that the cation levels used in these experiments were too high to observe the specificity towards K^+ . In support of this view was the finding by Humble and Hsiao (1969) that, at 100 mM cation concentrations, wide stomatal opening occurred in epidermal strips of *V. faba* even in the dark. Nevertheless, Travis and Mansfield (1979) find that when epidermal strips of

Figures 7.1–7.6 Histochemical tests for K^+ in the epidermis of various species with open or closed stomata. The test closes open stomata. Blackened areas indicate detection of K^+. Figures 7.1 and 7.2: upper epidermis from cotyledons of *Arachis hypogea* (peanut) with open and closed stomata, respectively. K^+ is located in the guard cells of open stomata. Figures 7.3–7.5: lower epidermis of *Avena sativa* (oat) showing the movement of K^+ from the subsidiary cells when the stomata are closed to the guard cells when the stomata are open. Figure 7.6, K^+ storage in the epidermal cells of *Bryophyllum tubiflorum*.

C. communis are indicated in 50 mM KCl stomatal responses to CO_2 and light and dark are similar to those observed in intact leaves. Fifty mM KCl may still be a non-physiological apoplastic concentration since the total concentration of ions in the leaf apoplast has been estimated to be only a few mM. However, the ion concentration in the apoplast of cells surrounding stomata may be much higher than elsewhere in the plant. Also, the concentration of K^+ and certain other ions is much higher in the epidermis than the mesophyll of the leaf. Additionally, it has already been noted that in *Cakile maritima*, a facultative halophyte, Na^+ accumulates in guard cells of plants growing in a saline medium.

Depending on the concentration, Ca^{++} reduces or completely suppresses K-stimulated stomatal opening in epidermal strips of a number of species (see Meidner and Willmer 1975). Pallaghy (1970) also found that, depending on the Na^+ or K^+ concentration, Ca^{++} could determine the Na^+ or K^+ selectivity for stimulating stomatal opening in epidermal strips of *V. faba*.

Although the specificity of the histochemical tests for Ca^{++} used by Fujino (1967) were dubious (particularly his $AgNO_3$ test which detected Cl^- amongst other things) results indicated that Ca^{++} increased in guard cells during stomatal opening and EDTA, a chelator of Ca^{++}, enhanced stomatal opening in epidermis of *C. communis*. Whether the effect of Ca^{++} is specific to guard cells or is part of a more general one characteristic of divalent cations needs further investigation.

Magnesium is reported to have no effect or cause slight suppression of K^+-stimulated opening in epidermal strips of a variety of species.

Thus, stomata may respond differently to the cations depending on the concentration, the ion species and whether Ca^{++} is present or absent. There may also be species differences in stomatal responses to the different ions and the stomata in epidermal strips of some species are not stimulated to open by any ion species or ion concentration.

7.2c Anions

The relatively high concentration of K^+ (or Na^+) which enters guard cells upon stomatal opening must be balanced by an equivalent negative charge. This is achieved in guard cells by either uptake of an equivalent amount of Cl^- or by uptake of some Cl^- plus some H^+ excretion and formation of organic acid anions.

Using epidermal strips of a number of species Cl^-, Br^- and NO_3^- salts of K^+ will stimulate opening to similar extents but $K_2SO_4^{--}$ is less effective at stimulating opening. Humble and Raschke (1971) concluded that in *V. faba*, SO_4^{--} and P_i were not involved in ionic balancing since microprobe data did not indicate an accumulation of P or S when stomata opened. However, the microprobe will only detect the absolute amount of S or P within the cells and it is possible that there may be binding and releasing of these elements within the guard cell during stomatal movements. Indeed, Fujino (1967) noted that there was an increase in the amount of inorganic phosphate in guard cells of closed stomata compared to open stomata (Fig. 7.7).

There is much evidence, however, indicating that Cl^- is the major inorganic anion to accumulate in guard cells to assist in the balance of charge (Fig. 7.8).

The extent to which Cl^- acts as the counterion for K^+ varies greatly, even within the same species, according to the source of the information. At one time it was considered that there may be species differences in the $K^+:Cl^-$ ratios of guard cells particularly since, in early studies, it was found that a small proportion of K^+ was balanced by Cl^- in *V. faba* while in maize a considerable amount of Cl^- acted as the counterion for K^+. However, the availability of Cl^- and K^+ to the plant and hence the guard cells seems much more of a determinant and may explain the wide variation of $K^+:Cl^-$ ratios observed in *V. faba* guard cells.

In *Allium cepa*, and possibly other species with starch-free guard cells, it has been reported that there is an absolute requirement for Cl^- as the counterion for K^+ (Raschke 1979).

The excess K^+ which is not balanced by Cl^- is considered to be neutralised by H^+ efflux (or OH^- uptake) and malate formation by the guard cells. If Cl^- is not available, H^+ efflux may balance the whole of K^+ uptake by the guard cells of *V. faba* (Humble and Raschke 1971).

The source of the protons is in the formation of organic acids from carbohydrates within the guard cells. Malic acid synthesis may be the chief source of the protons since numerous investigators have confirmed the observation of Allaway (1973) that malate accumulates in guard cells as stomata opened. It is possible that malate and Cl^- complement each other in balancing K^+ in guard cells, i.e. protons from malate synthesis will balance any excess K^+ not balanced by Cl^-. However, this may not always be the case since Allaway (1981) found that the presence of Cl^- did not always prevent increases of malate in guard cells of *V. faba* when stomata opened. Moreover, Outlaw and Lowry (1977) found that not only was citric acid the most abundant acid in guard cells of *V. faba* but, when stomata were open, the citrate concentration increased by about times a half while the malate concentration increased by about three times. If the pH of the cytoplasm is about 7, citrate (with pK values for the carboxyl groups of 3.1, 4.8 and 6.4) could be a source of more protons than malic acid (with pK values of 3.4 and 5.3). Although increased citrate levels were not found in 'rolled' epidermis of *V. faba* (epidermal cells are

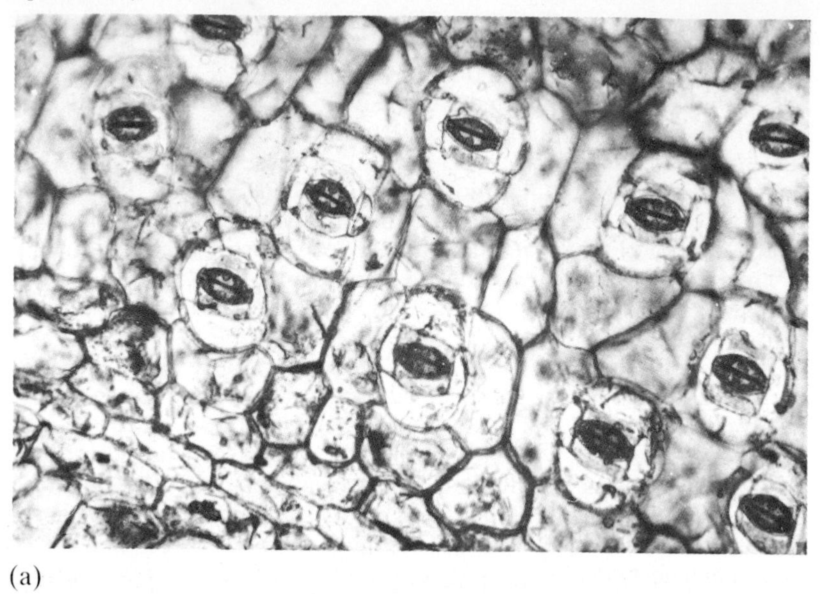

(a)

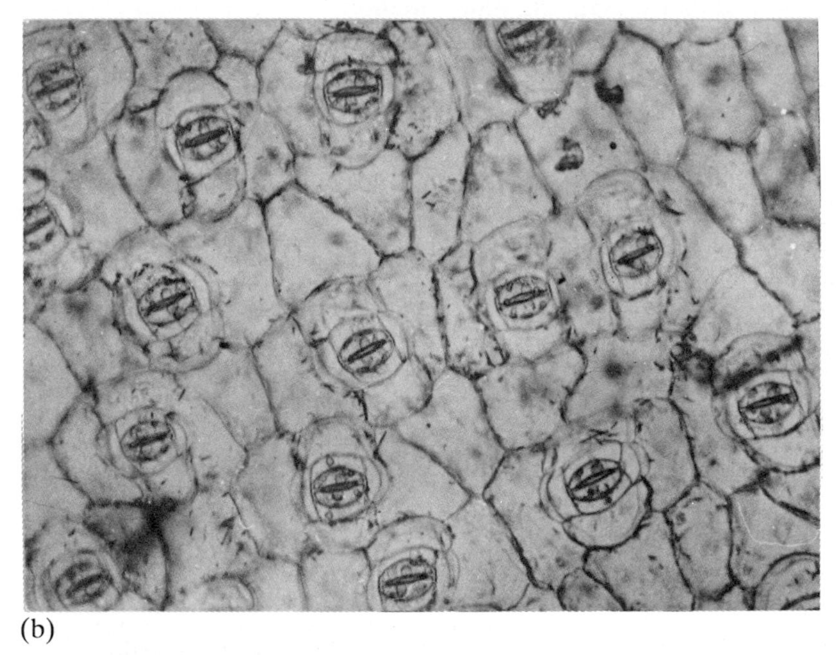

(b)

Figure 7.7 Histochemical test for inorganic phosphate in the epidermis of *Commelina communis* with (a) closed stomata and (b) open stomata (the test causes almost complete closure of open stomata). Inorganic phosphate is detected in the guard cells of closed stomata, indicated by an intense blue colouration (darkening in the photograph), but is not detected in guard cells of open stomata.

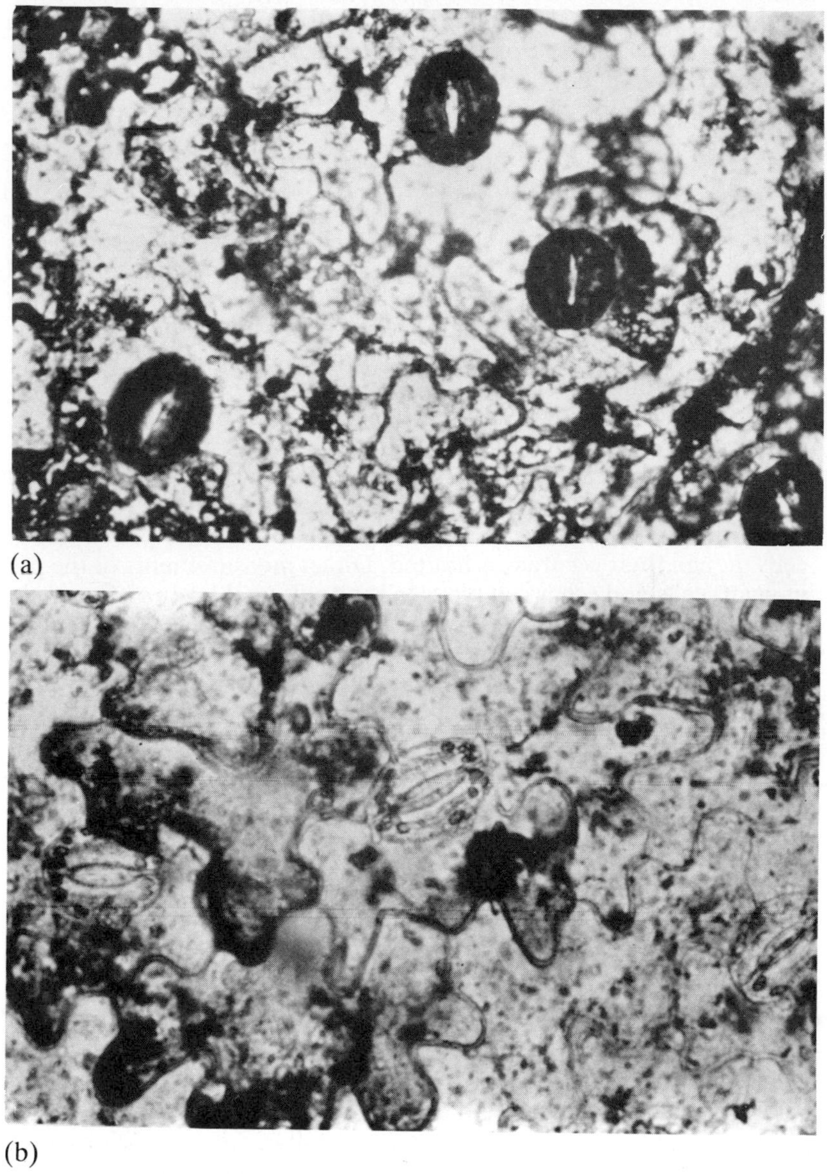

(a)

(b)

Figure 7.8 Histochemical test for Cl^- in epidermal strips of *Vicia faba* with (a) open stomata and (b) closed stomata (the test closes open stomata). High levels of Cl^-, indicated by blackened areas, appear in guard cells from open stomata while it is virtually absent in guard cells of closed stomata and more apparent in epidermal cells.

preferentially burst, leaving intact guard cells) with open stomata (Allaway 1981), in view of the importance of the information to our understanding of how stomata function more attention should be paid to the role of citrate.

7.2d pH values of guard, subsidiary and epidermal cells during stomatal movements

Table 7.2 shows the large changes of pH values which occur in the different cell types of the epidermal layer when stomata open and close. Values of about 4.0 to 5.5 and 5.5 to 6.5 are recorded for guard cells of closed and open stomata, respectively. pH values, however, will vary with changing stomatal apertures and possibly there will be species differences. Even larger pH changes are observed in epidermal and subsidiary cells as stomata open and close with the changes moving in the opposite direction to those in the guard cells. This suggests that there is a proton shuttle between guard cells and neighbouring cells during stomatal movements.

pH indicator dyes have been mainly used to measure pH values of cells though their accuracy is limited. Direct measurements of the pH in guard cells and other cell types in the epidermal layer of *Commelina communis* were made by Penny and Bowling (1975) using micro-pH electrodes implanted into the cells. They found values for guard cells of open and closed stomata to be 5.6 and 5.2 respectively.

The pH values being determined are probably vacuolar values and not cytoplasmic ones. Cytoplasmic pH values may remain fairly constant during opening and closing of stomata (possibly due to a pH 'stat' mechanism) while vacuolar pH values are known to vary in some species (e.g. in green cells of CAM plants). The primary process leading to the pH changes will be the fluxes of protons (or OH^- ions) but the exact control of the fluxes, particularly at the location of the tonoplast and plasmalemma, is not understood.

7.2e Electrophysiology of guard cells

Measurements of electrical properties of guard cells can assist in our understanding of how stomata function, particularly at the membrane level. However, considerable difficulties are encountered with such studies since the walls of guard cells are very thick and tough, making insertion of glass micropipettes (with tip diameter in the order of 1 μm) difficult. Additionally, guard cells are relatively small cells with small vacuolar volumes; the vacuoles may also consist of an anastomosing network of tubules as in *Allium cepa*. These features make it difficult to determine if the tip of the micropipette is located in the cytoplasm or vacuole of the guard cell.

Numerous groups have measured electrical potentials of guard cells (usually it has been concluded or assumed that the micropipette

Table 7.2. Changes of pH values (presumed vacuolar) in different cell types of the epidermal layer when stomata are open and closed

Species	Cell type	Open stomata*	Closed stomata	ΔpH (relative to cells when stomata are open)	Reference
Commelina communis	Guard cell	5.60	5.19	0.41	Penny and Bowling (1975)
	Outer lateral Subsidiary cell	5.56	5.78	−0.22	
	Epidermal cell	5.11	5.74	−0.63	
	Inner lateral Subsidiary cell	not measured	5.60	—	
Rumex acetosa	Guard cell	5.6	4.5	1.1	Pekarek (1936)
	Epidermal cell	5.6	7.7	−2.1	
Rumex acetosa	Guard cell	6.3	5.7	0.6	Pallas (1966)
Rumex acetosa	Guard cell	6.7	5.0	1.7	Scarth (1929)
Ophioglossum	Guard cell	6.4	5.6	0.8	Dayanandan and Kaufman (1975)
	Subsidiary cell	5.8	6.9	−1.1	
Zebrina pendula, *Tradescantia* spp.	Guard cell	6.0–7.4	<5.0	—	Scarth (1932)
	Epidermal cell	no pH difference detectable		—	

* Apertures are not given for open stomata. One must be aware, however, that the pH values for the guard cells (and probably other cell types) may vary according to the extent of stomatal opening, larger pH differences occurring as stomatal apertures increase.

tips are recording from the vacuole) and a wide variation in values has been obtained. This wide variation may be due not only to the problems indicated above, but also to species differences, whether epidermal strips, paradermal sections, isolated cells or whole leaves or shoots are used as the experimental material, and upon the external medium from which recordings are taken (potentials will vary according to the ionic content of the medium bathing the cells; if whole leaves are used the medium supplied to the leaves and from which recordings are taken may be considerably different from the apoplastic medium surrounding the guard cells). However, in epidermal strips or paradermal sections, guard cell vacuolar potentials are more negative than subsidiary or epidermal cell vacuolar potentials in the light or when stomata open; in the dark or when stomata close vacuolar potentials of guard cells depolarise to values approaching those of subsidiary or epidermal cells. Table 7.3 gives some values recently obtained of vacuolar potentials of guard, subsidiary and epidermal cells of various species.

The changes in vacuolar electrical potential brought about by light and dark or when stomata are open and closed may be a consequence of ion fluxes which are occurring during stomatal movements. Moody and Zeiger (1978) consider that the relatively fast changes in electrical potentials in response to light and dark and the sign and magnitude of the change suggest a light-sensitive proton pump which could provide the driving force for ion transport across the cell. Saftner and Raschke (1981), however, concluded that the electrical potentials of guard cells was a diffusion potential since, even if the concentration of alkali cations in the external medium was changed (between 3 and 300 mM), the vacuolar potential responded in a manner which agreed with the Nernst equation. These observations are to be expected if protons are actively pumped out of the guard cells and K^+ exchanges for them.

Electrical potential differences between cell types in the epidermal layer have also been measured. Using *Commelina communis*, Penny and Bowling (1974) found values near zero which did not appreciably differ between the different cell types whether stomata were open or closed. Penny and Bowling (1974) and Penny *et al.* (1976) have also measured K^+ and Cl^- concentrations in the various cell types when stomata were open and closed and concluded that K^+ and Cl^- transport to and from the guard cells was an active process; K^+ and Cl^- transport between subsidiary cells also appeared to be active. Caution is needed in the interpretation of the results, however, since invariably guard cells of open stomata collapsed.

Saftner and Raschke (1981) have also investigated the electrical properties of guard cell walls of *Commelina communis*. They find that the wall developed a potential of between −60 and −70 mV in 30 mM KCl at pH 7.0 and that the potential was not affected by

Table 7.3 Vacuolar potentials of epidermal, subsidiary and guard cells in epidermal strips of various species (after Saftner and Raschke 1981)

Species	Incubation medium	Stomatal aperture	Electrical potential (mV)			
			Epidermal cell	Subsidiary cell		Guard cell
				inner later	outer later	
Allium cepa		8 μm	−26			−39
Commelina communis	30 mM KCl	12–14 μm	−28	−30	−29	−45
	in 10 mM	closed	−32	−30	−31	−30
Nicotiana glauca	MOPS	8 μm	−23			−50
Vicia faba	buffer, pH 7.0	13 μm	−27			−42
Zea mays		8 μm	−29			−50

inhibitors, detergents or freezing. They concluded that guard cell walls behaved like Donnan spaces with fixed negative charges at a density of between 0.3 and 0.5 M. Also, they found that the change in pattern of the charge density to changing pH of the external medium was identical to the titration curve of polygalacturonic acid consistent with the view that guard cell walls are pectinaceous.

It is apparent that a relatively confusing picture emerges from electrophysiological studies of guard cells and further studies are needed to clarify some of the issues.

7.2f The 'malate-switch' hypothesis

An interesting 'malate-switch' hypothesis attempting to explain aspects of ion transport into and out of guard cells was proposed by Bowling (1976). The hypothesis is based upon the change of ionisation of malate with pH and assumes that the monovalent (H malate^{1-}) form is able to move out of cells while the divalent (malate^{2-}) form is not. Bowling observed in *Commelina communis* that a gradient of vacuolar pH occurred across the epidermal, subsidiary and guard cells, being highest in guard cells (5.8) and lowest in epidermal cells (5.1) when stomata were open and in the reverse direction when stomata were closed. At pH 5.1 malate is about equally ionised into the monovalent and divalent forms while at pH 5.8 about 90 per cent is in the divalent form, the remainder being in the monovalent form. If the pH changes within the various cell types precede the opening and closing movements then this could result in the following situation. When stomata close, guard cell pH values initially fall (become more acidic), resulting in the formation of more monovalent malate which can be released. At the same time the epidermal and subsidiary cell pH values increase (become less acidic) and, as the monovalent form enters these cells, it is converted to the divalent form and withheld. The reverse situation would occur upon opening. This corresponds with observations that malate levels in guard cells fall as stomata close and increase as stomata open.

The hypothesis has been criticised for a number of reasons. First, there is evidence that much of the malate metabolism is restricted to the guard cells and that during opening the carbon skeleton for malate synthesis is derived from starch while, upon closure, carbon from malate is channelled back into starch (*see* §7.3). Second, the pH changes which occur in guard cells indicate that the levels of the malate^{2-} form is equal to or greatly more than the H malate^{1-} form and, therefore, the bulk of the malate would not be free to move from cell to cell. Third, the hypothesis assumes that malate fluxes follow K^+ fluxes during stomatal movements and, although malate efflux from the guard cells has been observed when stomata close, influx of malate has not been recorded when stomata open. Fourth, stomata may be able to function without the participation of malate acting as a counterion for K^+. In *Allium cepa*, for example, it has been suggested that Cl^- brings about complete charge balance for K^+. A fifth problem is that the hypothesis refers to vacuolar pH values and ignores the cytoplasmic ones. Once the malate enters the cytoplasm with a pH assumed to be about 7.0 then virtually all the malate will be in the malate^{2-} form and will be trapped. However, the changing ionisation of malate may still be important not in controlling malate fluxes into and out of the guard cells but in controlling fluxes between the cytoplasm and vacuole of the guard cell.

7.3 The metabolism of guard cells

This section concentrates on the carbon metabolism of guard cells paying particular attention to the carbon flow between starch and malate during stomatal movements. For further reading the reader is referred to recent reviews by Raschke (1979), Milthorpe *et al.* (1979) and Willmer (1981).

7.3a The source of malate

Willmer *et al.* (1973) detected high levels of PEP carboxylase, malate dehydrogenase (NAD and NADP specific) and NADP malic enzyme activity in epidermal tissue of tulip and *C. communis*. Subsequently, other workers detected considerable activity of enzymes involved in malate metabolism in epidermal tissue of a variety of species. Although there is some evidence that the enzymes were located primarily in the guard cells of the epidermis the results of Outlaw and his colleagues (*see* Outlaw 1980) are unequivocal in this respect since they measured activities in isolated single guard cell pairs. They found that PEP carboxylase activity in guard cells of *V. faba* is much more than sufficient to account for malate accumulation necessary to balance all K^+ which would accumulate in guard cells of open stomata. Table 7.4 presents some of their results obtained for single guard cell pairs of *V. faba*.

Such enzyme activities suggest that malate can be synthesised and decarboxylated in the following series of reactions:

$$CO_2 + PEP \xrightarrow{\text{PEP carboxylase}} OAA \xrightarrow[\text{NAD(P)H}_2 \;\to\; \text{NAD(P)}]{\text{Malate dehydrogenase}} MALATE \xrightarrow[\text{NAD(P)} \;\to\; \text{NAD(P)H}_2]{\text{Malic enzyme}} PYRUVATE + CO_2$$

OAA ↓ (GLUTAMATE → ; aspartate aminotransferase ; → OXOGLUTARATE) ↓ ASPARTATE

Table 7.4 Enzyme activities (μ moles substrate converted/h/mg chlorophyll or protein as indicated) in guard cells and other cell types of *Vicia faba* L. leaves as determined by quantitative histochemical techniques (from Outlaw 1982, and Outlaw, private communication)

Enzyme	Cell type							
	Guard cell		Epidermal cell		Palisade cell		Spongy mesophyll cell	
	Protein	Cholorophyll	Protein	Chlorophyll	Protein	Chlorophyll	Protein	Chlorophyll
RuBP carboxylase (EC 4.1.1.39)	<0.1	<4.0	<0.5	—	5.5	110	5.0	202
P-ribulokinase (EC 2.7.1.19)	<5.0	<175	<2.4	—	82	1,632	50	1,994
NADP glyceraldehyde-P-dehydrogenase (EC 1.2.1.13)	<0.5	<18	<0.4	—	5.8	115	3.2	130
PEP carboxylase (EC 4.1.1.31)	12.7	445	3.9	—	1.5	30	1.7	70
NADP malic enzyme (EC 1.1.1.40)	10.2	357	12.5	—	0.5	9.6	2.8	113
NAD malic enzyme (EC 1.1.1.38)	40	1,416	16.2	—	5.0	100	7.1	283
PEP carboxykinase (EC 4.1.1.32)	<1.3	<47	—	—	—	—	—	—
Pyruvate P_i dikinase (EC 2.7.9.1)	<0.4	<14	ND	—	ND	ND	ND	ND

ND = non-detectable

^{14}C labelling studies also suggest that the above steps occur in guard cells. When $^{14}CO_2$ or $NaH^{14}CO_3$ is supplied to epidermal tissue over short-term experiments, generally, malate is chiefly labelled with lesser amounts of label in aspartate, some other amino acids, acids of the TCA cycle (other than malate) and sugars and sugar phosphates (*see* Milthorpe *et al.* 1979; Willmer 1981). Some investigators do not detect label in sugars and sugar phosphates and Raschke and Dittrich (1977) believe that the presence of label in these compounds is due to contamination of the epidermis by mesophyll cells or chloroplasts and represents photosynthetic products.

A further divergence of opinion concerns ^{14}C labelling patterns observed when isolated protoplasts of *V. faba* guard cells are exposed to $^{14}CO_2$. Schnabl (1980) finds that label is channelled only into malate, aspartate, glycollate and starch in the light and the dark. Another study, however, finds label in malate, aspartate, sugars, starch, a variety of other amino acids and sugar phosphates but, notably, very little in glycollate; there was also a considerable light enhancement of the CO_2 assimilation rate (Brown, Cohen and Outlaw, private comm.). The greatly increased CO_2 fixation in the light by guard cell protoplasts is particularly interesting in view of the finding that there is only a 1–3 fold light enhancement in epidermal strips. The rates of CO_2 assimilation by guard cell protoplasts or epidermal strips must also be put into focus: they are 100 to 1000 times less than in mesophyll cells.

^{14}C labelling patterns and turnover patterns also differ according to the stage of stomatal movements and, although the significance of the differences are not entirely understood, presumably reflect the changes of direction of flow of carbon. Such studies and other investigations of sucrose levels in isolated guard cell pairs or epidermal tissue when stomata are open and closed indicate that there is a pool of sucrose in the guard cells which increases as stomata open and which has a fairly direct and reversible access to carbon from starch and malic acid.

Evidence indicates that starch supplies carbon skeletons for malate production during stomatal opening (except in starch-free guard cells where fructans may substitute for starch – *see* §7.3b). Moreover, starch is an ideal substance from which malate (a source of ions and osmoticum) can be made since it is non-ionic and essentially insoluble therefore having little osmotic activity. Most observations indicate that guard cell starch levels decrease when stomata open and increase when stomata close, although some investigators do not observe this inverse correlation. However, as Raschke (1979) and Allaway (1981) point out, if there is simultaneous consumption within the guard cell and replenishment of carbohydrate from outside then strict synchrony may not occur. Figure 7.9 depicts some of the major pathways which may be operating in guard cells during stomatal movements

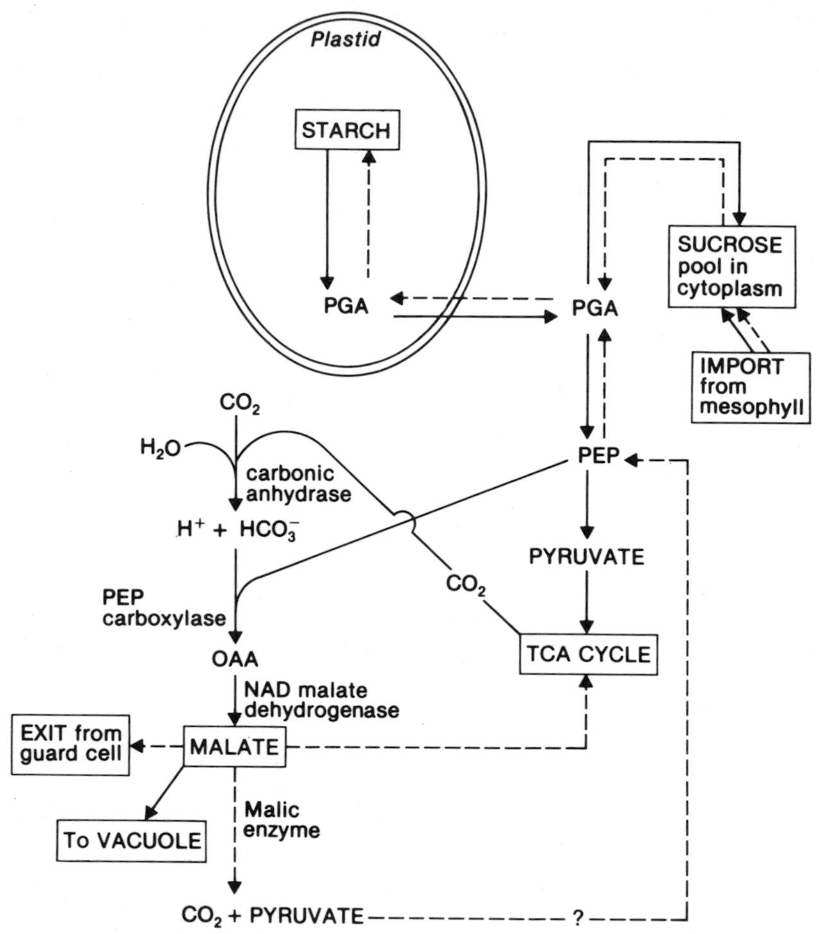

Figure 7.9 Possible key pathways which may operate in starch-containing guard cells during stomatal opening (——) and closing (----). Enzymes which have been detected in guard cells are shown catalysing appropriate reactions; high levels of activity of an enzyme for converting pyruvate to PEP has not yet been unequivocally detected.

and indicates the route of net carbon flow during stomatal opening and concomitant malate synthesis.

7.3b The role of chloroplasts in guard cells

In the early work of Willmer *et al.* (1973) high levels of RuBP carboxylase activity were detected in the epidermis of *C. communis* and tulip. Subsequently, it has been concluded that the Calvin cycle does not function in the epidermis of *C. communis* since phosphoribulokinase was not detectable (Raschke and Dittrich 1977). Additionally, they suggested that the high RuBP carboxylase activity found in the earlier study was due to contamination of the epidermis with mesophyll. Work by Outlaw *et al.* (1979) using single guard cell pairs dissected from freeze-dried leaves established the absence of the Calvin cycle in guard cells of *V. faba* and tobacco since they found no evidence of RuBP carboxylase activity and insignificant levels of phosphoribulokinase and NADP specific glyceraldehydephosphodehydrogenase activity. Additionally, in the same work, immunoelectrophoretic studies did not detect RuBP carboxylase in extracts from guard cell protoplasts of *V. faba*. Further work by Outlaw and his colleagues indicates that RuBP carboxylase activity is absent from guard cells of a number of other species (private communication). Immunofluorescence techniques, however, detected considerable amounts of RuBP carboxylase in guard cells of CAM plants, small amounts in some C_3 plants but none in C_4 plants.

The Calvin cycle will also be absent in guard cells of those species which lack chloroplasts, although starch-containing plastids are present, as in certain species of *Paphiopedilum* and in a variegated cultivar of *Pelargonium zonale*.

Thus, in some species at least, fixation of CO_2 via the Calvin cycle does not occur in guard cells. This may turn out to be a common feature of guard cells in most species. However, except for a few genera such as *Allium*, green and non-green plastids of guard cells contain an abundance of starch. The plastids may, therefore, be considered as amyloplasts and, in general, as stomata open guard cell starch decreases while starch levels increase as stomata close.

In the starch-free chloroplasts a search was made for other soluble polysaccharides which could supply part of the carbon skeleton for malate synthesis in guard cells. Schnabl (*see* Schnabl and Ziegler 1977) concluded that the mucilaginous polysaccharides detectable in *Allium cepa* epidermis did not serve as a carbon source for malate synthesis and that fructans or other soluble polysaccharides which could substitute for starch were not detectable. Darbyshire and Allaway (in Allaway 1981), however, found considerable amounts of fructans (polymerised to a degree of about 8 or 9 hexose units) in dark-treated epidermis of *Allium cepa* (in which guard cells were the only intact cells), most of which disappeared when the epidermis was given a light-treatment. Thus, these investigators believe that fructans could be mobilised to supply carbon for malate synthesis in guard cells of *A. cepa*.

Little is currently known about the enzyme complement of guard cell plastids though it appears that those necessary for the interconversion of starch and PGA and/or triose phosphate must be present since evidence indicates that there is a flow of carbon between starch and malic acid during stomatal movements and PGA and/or triose phosphate are compounds produced from starch hydrolysis which are able to readily cross the chloroplast envelope (assuming the envelopes of guard cell chloroplasts have similar properties to mesophyll chloroplasts).

Information on the light reactions of guard cell chloroplasts is, at present contradictory.

Chlorophyll a/b ratios for guard cells indicate whether their chloroplasts are enriched in photosystem (PS) I or II; values approximately between 2 and 3 (typical of C_3 plants) indicate that considerable amounts of both photosystems are present while values above 3 (as in C_4 plants) indicate that PS I dominates. Many species have been investigated for their chlorophyll a/b ratios by a number of independent investigators and a large variation has been obtained spanning C_3 and C_4 values. Most investigators analysed epidermal tissue to obtain chlorophyll a/b ratios of guard cells. Unfortunately, epidermal tissue is not ideal material for making such determinations due to possible contamination from underlying mesophyll cells and the very low content of chlorophyll in guard cells making accurate measurements difficult. However, Yemm and Willis (1954) used microspectrophotometric analysis of single plastids of guard cells in *Rumex patientia*, *Chrysanthemum maximum* and *V. faba* and found values of 2–3. More recently Outlaw *et al.* (1980) obtained a value of 2.8 for protoplasts of *V. faba* guard cells.

Other more sophisticated methods have been used to determine which photosystems are present in guard cell protoplasts. Delayed light emission, P_{700} content and absorption and low temperature fluorescence emission spectra of protoplasts of *V. faba* guard cells indicate that both photosystem I and II are present (Outlaw *et al.* 1980). Low temperature emission spectra and variable fluorescence induction kinetics of white tissue from leaves of *Chlorophyllum comosum* (in which guard cells were the only cells to contain chloroplasts) also indicated that both PS I and II were present and that PS II was probably larger in guard cell chloroplasts than in mesophyll chloroplasts (Ziegler *et al.* 1980). This study also concluded that guard cell chloroplasts oxidise water and have a functional PS II connected through an intermediate plastoquinone pool to PS I resulting in the generation of reduced NADP, i.e. a linear (non-cyclic) electron transport mechanism was present.

Schnabl and Hampp (1980), however, could only detect PS I in protoplasts of *V. faba* guard cells as determined by O_2 consumption or evolution in the light in the presence of a variety of electron acceptors and donors.

Observations of stomatal movements in epidermal strips of a number of species in the presence of inhibitors of PS I or PS II or in red or far-red light also suggested that PS I and the accompanying cyclic photophosphorylation was necessary for opening to occur.

Whether both photosystems occur in chloroplasts of guard cells or just the cyclic system their activities will be small relative to that occurring in chloroplasts of mesophyll cells and, therefore, may not be of great significance. Nevertheless, as pointed out by Outlaw *et al.* (1980), if non-cyclic electron flow occurs in chloroplasts of guard cells while triose-P-dehydrogenase (which assists in catalysing the reductive step, PGA → phosphoglyceraldehyde) is absent then the purpose of the $NADPH_2$, produced by non-cyclic electron flow, remains obscure. (In other green cells of higher plants which possess PS I and II the step, PGA to phosphoglyceraldehyde, which utilises NADPH produced in linear electron flow is also present.)

Presumably ATP produced in the chloroplasts of guard cells can be shuttled out and used for ion transport processes occurring at the plasmalemma and tonoplast.

7.3c Carbon intake by guard cells

On the assumption that the stores of guard cell starch are not sufficient to satisfy the 'life-time' demands for carbon to make malate and as a source of energy in the TCA cycle, and if the Calvin cycle is not present in guard cells, then import of carbon by the cells must occur. (The Calvin cycle is an autocatalytic process being able to regenerate substrate to perpetuate the cycle and also allow carbon to be syphoned off for sucrose synthesis.)

In some species (e.g. *Phyllitis scolopendrium* and other ferns) epidermal cells possess an abundance of chloroplasts and there is some evidence indicating that they can photosynthesise and may, therefore, supply guard cells with carbon. However, epidermal cells are often free of chloroplasts and in such cases the guard and epidermal cells must receive carbon from mesophyll cells. There is considerable experimental evidence to suggest this is the case (*see* Milthorpe *et al.* 1979; Willmer 1981).

A particularly convincing piece of evidence comes from the work of Outlaw and Fisher (1975), who showed that, following a pulse of $^{14}CO_2$ to a leaflet of *Vicia faba*, label eventually reached the epidermal layers. Also, the observation that exogenously supplied sugars and glucose-1-phosphate to epidermal tissue of numerous species resulted in starch formation within the guard cells is a further indication that metabolite transport could occur between mesophyll cells and guard cells.

7.3d The fate of malate

When stomata close, malate levels in guard cells decrease. Three fates of malate have been suggested: (1) malate is released from the guard cells; (2) malate is metabolised in the mitochondria of the guard cells; and (3) malate can be channelled back into starch synthesis within the guard cell (see Fig. 7.9). All three possibilities probably occur in guard cells although the extent to which they occur is not known.

Van Kirk and Raschke (1978) concluded that a substantial proportion of malate was released from guard cells to the apoplast when stomata closed. Also, Dittrich and Raschke (1977) found that only labelled malate was released from epidermal tissue of *C. communis* with closing stomata which was exposed to $^{14}CO_2$. However, other studies have shown that numerous labelled compounds are released from epidermal tissue of *C. cyanea* at all stages of stomatal movements (see Milthorpe *et al.* 1979; Willmer 1981) while Freer-Smith and Willmer (1981) found that malate was the chiefly labelled compound (about 90% of the total ^{14}C) released from epidermal strips of *C. communis* whether stomata were opening or closing. Thus, some caution is needed in the interpretation of such studies; the results may, for example, only reflect the extent to which cell damage with accompanying cell 'leakiness' is occurring.

Malate can also readily cross mitochondrial membranes and be used as a source of energy for active ion transport systems occurring in the guard cells or for general cell metabolism. The presence of labelled acids of the TCA cycle when epidermal tissue is fed $^{14}CO_2$ or ^{14}C-malate indicates this to be the case.

There is also evidence from ^{14}C labelling studies that part of the carbon from malate is channelled into starch synthesis via gluconeogenesis (the synthesis of sugars from organic acids) (Dittrich and Raschke 1977; see also Milthorpe *et al.* 1979; Willmer 1981). Malic enzyme activity has been detected in epidermal tissue of a variety of species and guard cells of *V. faba* and, therefore, decarboxylation of malate can occur. However, the resulting pyruvate must first be converted to PEP before it can be channelled into gluconeogenesis since the step, pyruvate to PEP, is essentially irreversible. The search for PEP synthesising enzymes (pyruvate P_i dikinase and PEP carboxykinase) by numerous independent groups has resulted in conflicting reports; one group finds high levels of activity of both enzymes in the epidermis of *Tridax procumbens* and *Commelina benghalensis* while other groups detect little or no activity.

7.3e The energy source for stomatal movements

Both opening and closing movements have been described as active, energy-consuming steps. The opening movement can be more readily conceived of as being active since guard cells are at their lowest pressure (in a 'relaxed' state) when stomata are closed and increase their pressure as stomata open. Also, a number of studies have shown that metabolic inhibitors (uncouplers of phosphorylation activity and inhibitors of electron flow) prevent stomatal opening (*see* §7.3b).

However, numerous studies have also shown that metabolic inhibitors, such as sodium azide, which is an inhibitor of respiration, and oxygen-free air prevent stomatal closure. This indicates that closing is an active process.

Guard cells contain an abundance of mitochondria while, in most plants, chloroplasts are poorly developed which suggests that oxidative phosphorylation is the major source of ATP for 'active' stomatal movements. Moreover stomata can open in the dark (and do as a normal course of events in CAM plants) and stomatal closing, which may be active, is brought about in darkness, conditions under which photophosphorylation will not be occurring. Under these situations the energy source must be from oxidative phosphorylation or a system other than photo-phosphorylation. Some inhibitor studies have also shown that, in the light, stomatal opening is dependent on cyclic photophosphorylation.

7.4 Summary of events occurring in guard cells during stomatal movements

Environmental factors such as light or low CO_2 concentrations trigger off events which may result in stomatal opening. How these signals are sensed and how they are transduced into driving the ion fluxes which control stomatal movements is not understood. However, a pigment, possibly a flavin or flavoprotein situated in or on the guard cell plasmalemma or tonoplast, may be a sensor of blue light and initiate H^+ excretion. Low CO_2 levels may also influence H^+ levels in the guard cell cytoplasm, resulting in H^+ excretion.

The major events which occur in guard cells when stomata open or close are illustrated in Fig. 7.10. The sequence of events is not clear

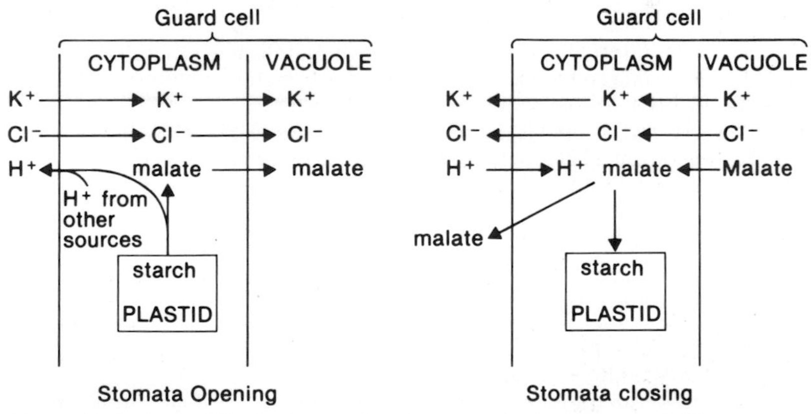

Figure 7.10 A simplified scheme showing the principal events which occur when stomata open and close.

but the following is suggested. When stomata open, protons are first pumped from the guard cell, resulting in hyperpolarisation of the plasmalemma. K^+ may then enter the guard cell passively. Some Cl^- also enters the guard cells but complete charge balance of the excess K^+ is accomplished by synthesis of malate. At cytoplasmic pH values, assumed to be about 7.0, malate will be ionised almost completely into the divalent (malate^{2-}) form which acts as a counterion for K^+. The carbon skeleton for malate synthesis is derived from starch. K^+, Cl^- and malate accumulate in the guard cell vacuole and are mainly responsible for the decreases in guard cell osmotic potential when stomata open.

Less is known about the initiation and control of stomatal closing but, essentially, the ionic migrations described above are reversed. Part of the carbon from malate is channelled into starch synthesis via gluconeogenesis and some malate may also leave the guard cell.

Further reading and references

Allaway, W. G. (1973) Accumulation of malate in guard cells of *Vicia faba* during stomatal opening, *Planta*, **110**, 63–70.

Allaway, W. G. (1981) Anions in stomatal operation, in Jarvis, P. G. and Mansfield, T. A. (eds), *Stomatal Physiology*. Cambridge University Press.

Allaway, W. G. and Hsiao, T. C. (1973) Preparation of rolled epidermis of *Vicia faba* L. so that stomata are the only viable cells: analysis of guard cell potassium by flame photometry, *Austr. J. Biol. Sci.*, **26**, 309–18.

Bowling, D. J. F. (1976) Malate-switch hypothesis to explain the action of stomata, *Nature*, **262**, 393–4.

Dayanandan, P., and Kaufman, P. B. (1975) Stomatal movements associated with potassium fluxes, *Amer. J. Bot.*, **62**, 221–31.

Dittrich, P. and Raschke, K. (1977) Malate metabolism in isolated epidermis of *Commelina communis* L. in relation to stomatal functioning, *Planta*, **134**, 77–81.

Eshel, A., Waisel, Y. and Ramati, A. (1974) The role of sodium in stomatal movements of a halophyte: a study by x-ray microanalysis, in Wehrmann, J. (ed.), *7th Intl. Colloq. on Plant Analysis and Fertilizer Problems.*, German Soc. Plant Nutrition. Hannover, Germany.

Freer-Smith, P. and Willmer, C. M. (1981) Guard cell metabolism in epidermis of *Commelina communis* L. during stomatal opening and closing, *J. Exp. Bot.*, **32**, 535–43.

Fischer, R. A. (1968) Stomatal opening: role of potassium uptake by guard cells, *Science*, **160**, 784–5.

Fujino, M. (1967) Role of adenosinetriphosphate and adenosinetriphosphatase in stomatal movement, *Sci. Bull. Fac. Educ. Nagasaki Univ.*, **18**, 1–47.

Heath, O. V. S. and Mansfield, T. A. (1969) The movement of stomata, in Wilkins, M. (ed.) *Physiology of Plant Growth and Development*, McGraw-Hill, London.

Humble, G. D. and Hsiao, T. C. (1969) Specific requirement of potassium for light-activated opening of stomata in epidermal strips, *Plant Physiol.*, **44**, 230–4.

Humble, G. D. and Raschke, K. (1971) Stomatal opening quantitatively related to potassium transport. Evidence from electron probe analysis, *Plant Physiol.*, **48**, 447–53.

Iljin, W. S. (1957) Drought resistance in plants and physiological processes, *Ann. Rev. Plant Physiol.*, **8**, 257–346.

Imamura, S. (1943) Untersuchungen uber den Mechanismus der Turgorschwankung der Spaltöffnungschlieszellen, *Jap. J. Bot.*, **12**, 251–346.

Kohl, F. G. (1895) Über assimilationsenergie und Spaltöffnungsmechanic, *Botanische Centralblatt*, **64**, 109–10.

Lloyd, F. E. (1908) The physiology of stomata, *Carnegie Inst.*, *Washington, Publ. No. 82*.

Macallum, A. B. (1905) On the distribution of potassium in animal and vegetable cells, *J. Physiol.*, **32**, 95–128.

Meidner, H. and Willmer, C. M. (1975) Mechanics and metabolism of guard cells, in *Curr. Advances in Plant Sci.*, no. 17.

Milthorpe, F. L., Thorpe, N. and Willmer. C. M. (1979) Stomatal metabolism – a current assessment of it features in *Commelina*, in Sen, D. N. (ed), *Structure, Function and Ecology of Stomata*. Bishen Singh and Mahendra Pal Singh, Dehra Dun, India.

Mohl, H. von (1856) Welche Ursachen bewirken Erweiterung und Verengung der Spaltöffnungen, *Bot. Ztg.*, **14**, 697–704, 713–21.

Moody, W. and Zeiger, E. (1978) Electrophysiological properties of onion guard cells, *Planta*, **139**, 159–65.

Nelson, S. D. and Mayo, J. M. (1979) Low K^+ in *Paphiopedilum leeanum* leaf epidermis: implications for stomatal functioning, *Can. J. Bot.*, **55**, 489–95.

Outlaw, W. H. Jr (1980) A descriptive evaluation of quantitative histochemical methods based on pyridine nucleotides, *Ann. Rev. Plant Physiol.*, **31**, 299–311.

Outlaw, W. H. Jr (1982) Carbon metabolism in guard cells, in Creasy, L. L., and Hrazdina, G. (eds), *Cellular and Subcellular Localization in Plant Metabolism*. Plenum, London.

Outlaw, W. H. Jr and Fisher, D. B. (1975) Compartmentation in *Vicia faba* leaves. 1. Kinetics of ^{14}C in the tissues following pulse labelling, *Plant Physiol.*, **55**, 699–703.

Outlaw, W. H. and Lowry, O. H. (1977) Organic acid and potassium accumulation in guard cells during stomatal opening. *Proc. Natl. Acad. Sci., USA*, **74**, 4434–8.

Outlaw, W. H. Jr, Manchester, J., Dicamelli, C. A., Randall, D. P., Rapp, B. and Veith, G. M. (1979) Photosynthetic carbon reduction pathway is absent in chloroplasts of *Vicia faba* guard cells, *Proc. Natl. Acad. Sci., USA*, **76**, 6371–5.

Outlaw, W. H. Jr, Mayne, B. C., Zenger, V. E. and Manchester, J. (1980) Presence of both photosystems in guard cells of *Vicia faba* L. Implications for environmental signal processing, *Plant Physiol.*, **67**, 12–16.

Pallaghy, C. K. (1970) The effect of Ca^{2+} on the ion specificity of stomatal opening in epidermal strips of *Vicia faba*, *Z. Pflanzenphysiol.*, **62**, 58–62.

Pallas, J. E. Jr (1966) Mechanics of guard cell action, *Quart. Rev. Biol.*, **41**, 365–83.

Pekarek, J. (1936) Über die Aziditätsverhältnisse in den Epidermis und Schliesszellen bei *Rumex acetosa* im Licht und im Dunkeln, *Planta*, **21**, 419–46.

Penny, M. G. and Bowling, D. J. F. (1974) A study of potassium gradients in the epidermis of intact leaves of *Commelina communis* L. in relation to stomatal opening, *Planta*, **119**, 17–25.

Penny, M. G. and Bowling, D. J. F. (1975) Direct determination of pH in the stomatal complex of *Commelina*, *Planta*, **122**, 209–12.

Penny, M. G., Kelday, L. S. and Bowling, D. J. F. (1976) Active chloride transport in the leaf epidermis of *Commelina communis* in relation to stomatal activity, *Planta*, **130**, 291–4.

Raschke, K. (1975) Stomatal action, *Ann. Rev. Plant Physiol.*, **26**, 309–40.

Raschke, K. (1979) Movements of stomata, in Haupt, W. and Feinleib, M. E. (eds), Physiology of movements, *Encyclopedia of Plant Physiology* (New series, vol. 7). Springer-Verlag, Berlin.

Raschke, K. and Dittrich, P. (1977) [^{14}C]carbon-dioxide fixation by isolated leaf epidermis with stomata closed or open, *Planta*, **134**, 69–75.

Raschke, K. and Fellows, M. (1971) Stomatal movements in *Zea mays*: shuttle of potassium and chloride between guard cells and subsidiary cells, *Planta*, **101**, 296–316.

Saftner, R. A. and Raschke, K. (1981) Electrical potentials in stomatal complexes, *Plant Physiol.*, **67**, 1124–32.

Sawhney, B. L. and Zelitch, I. (1969) Direct determination of potassium ion accumulation in guard cells in relation to stomatal opening, *Plant Physiol.*, **44**, 1350–4.

Scarth, G. W. (1929) The influence of H-ion concentration on the turgor and movement of plant cells with special reference to stomatal behaviour, *Proc. Intl. Conf. Plant Sci.*, **2**, 1151–62.

Scarth, G. W. (1932) Mechanism of the action of light and other factors on stomatal movements, *Plant Physiol.*, **7**, 481–504.

Schnabl, H. (1980) CO_2 and malate metabolism in starch-containing and starch-lacking guard-cell protoplasts. *Planta*, **149**, 52–8.

Schnabl, H. and Hampp, R. (1980) Chloroplasts derived from *Vicia* guard cell protoplasts lack photosystem II and ferredoxin – NADP: oxidoreductase activity. Naturwissensch, **67**, 465–6.

Schnabl, H. and Ziegler, H. (1977) The mechanism of stomatal movement in *Allium cepa* L., *Planta*, **136**, 37–43.

Stevens, R. A. and Martin, E. S. (1977) New structure associated with stomatal complex of the fern *Polypodium vulgare*. Nature, **265**, 331–36.

Travis, A. J. and Mansfield, T. A. (1979) Stomatal responses to light and CO_2 are dependent on KCl concentration, *Plant Cell and Environment*, **2**, 319–23.

Van Kirk, C. A. and Raschke, K (1978) Release of malate from epidermal strips during stomatal closure, *Plant Physiol.*, **53**, 360–5

Willmer, C. M. (1981) Guard cell metabolism, in Jarvis, P. G. and Mansfield, T. A. (eds) *Stomatal Physiology*, Cambridge University Press.

Willmer, C. M. and Mansfield, T. A. (1969) A critical examination of the use of detached epidermis in studies of stomatal physiology, *New Phytol.*, **68**, 363–75.

Willmer, C. M., Pallas, J. E. Jr and Black, C. C. Jr (1973) Carbon dioxide metabolism in leaf epidermal tissue, *Plant Physiol.*, **52**, 448–52.

Yamashita, T. (1952) Influences of potassium supply upon various properties and movement of guard cell, *Sieboldia*, *Acta Biol.*, **1**, 51–70.

Yemm, E. W. and Willis, A. J. (1954) Chlorophyll and photosynthesis in stomatal guard cells, *Nature*, **173**, 726.

Zieger, E., Armond, P. and Melis, A. (1980) Fluorescence properties of guard cell chloroplasts. Evidence for linear electron transport and light-harvesting pigments of photosystem I and II, *Plant Physiol.*, **67**, 17–20.

Zelitch, I (1963) The control and mechanism of stomatal movements. In *Stomata and water relations in plants*. Bull. 664, Conn. Exp. St., Newhavenen, Connecticut, 18–36.

Zlotnikova, I. F., Gunar, I. I. and Panichkin, L. A. (1977) Measurement of intracellular potassium activity in *Tradescantia* leaf epidermal cells, *Izv. Timiryazev. S-Kh. Akad.*, **2**, 10–16.

8 Some methods of studying stomata

There are numerous experiments that can be carried out to study various aspects of stomatal behaviour and operation without the use of expensive equipment. Outlined below are some experiments of varying complexity which can be done in most undergraduate laboratories.

8.1 Measurement of stomatal apertures

A brief appraisal of some of the different methods available for measuring stomatal apertures is given here.

Silicone rubber impressions of the leaf surface provide a permanent record without damaging the leaf and are easy to make in the field. However, measurements are not too accurate and the method is only useful when species with relatively large stomata are being investigated.

Leaf infiltration techniques are now rarely used and are not recommended since the liquids used to infiltrate the leaves are poisonous and damage the leaf tissues.

Stomatal apertures can be measured directly with a microscope in epidermal strips (if the epidermis peels readily from the leaf) or in paradermal sections. In both cases leaves are damaged and, although accurate measurements can be made, the method is time consuming.

Lloyd's strips (Lloyd 1908) can be prepared by fixing the epidermis in alcohol and measuring apertures at one's own convenience. However, some distortion of the cells may occur, giving inaccurate measurements of the original apertures.

Porometers usually provide the most accurate and convenient way of recording stomatal movements. They measure the average stomatal resistance (or conductance) of perhaps many thousands of stomata simultaneously, and they do not damage the leaf.

Two basic types of porometers exist, namely diffusion porometers and viscous or mass flow porometers. Diffusion porometers measure

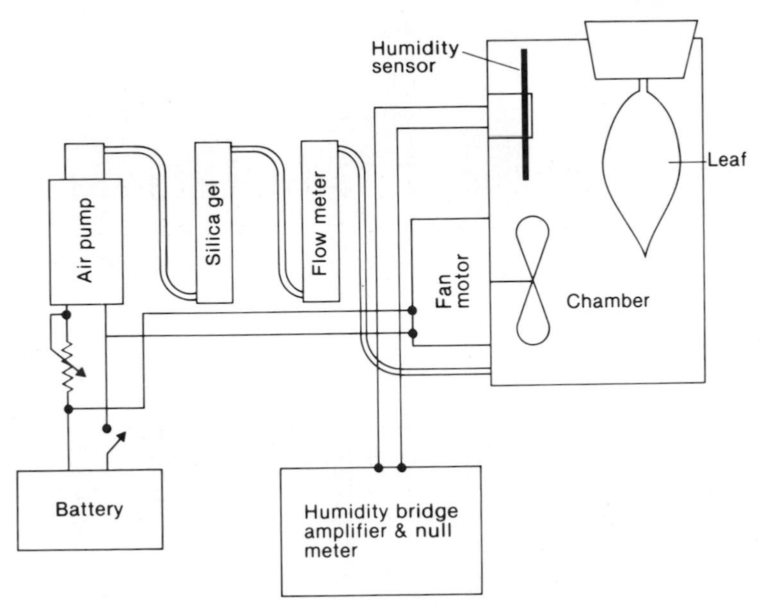

Figure 8.1 Block diagram of a steady-state diffusion porometer of the type designed by Beardsell *et al.* (1972) showing components and interconnections, (courtesy of Washington State University, College of Agriculture Research Centre).

the rate of diffusion of a gas across a leaf through the upper and lower leaf surfaces, as in the nitrous oxide diffusion porometer (Slatyer and Jarvis 1966), or of water vapour out of a leaf (Figs 8.1, 8.2). Viscous flow porometers measure the rate at which air can be pushed or sucked through a leaf when a pressure difference is applied across the leaf, i.e. they measure resistance to viscous flow (Fig. 8.3).

Porometers more correctly measure the resistance to gas movement across, or part way across, and out of a leaf (i.e. leaf resistance) and not just the resistance across the stomatal pore (i.e. stomatal resistance). However, if it is assumed that the resistance to gas movement across a leaf is a constant for the duration of an experiment then changes of leaf resistance are a reflection of changing stomatal apertures. It is, therefore, useful to obtain a relationship between the porometer resistance reading and the actual stomatal aperture.

The reader is referred to the book of Meidner and Mansfield (1968) and Bulletin 809, Washington State University (1975) for detailed descriptions of the different types of porometers with their advantages and disadvantages. However, a diffusion porometer of the type designed by Kanemasu *et al.* (1969) is commercially available and

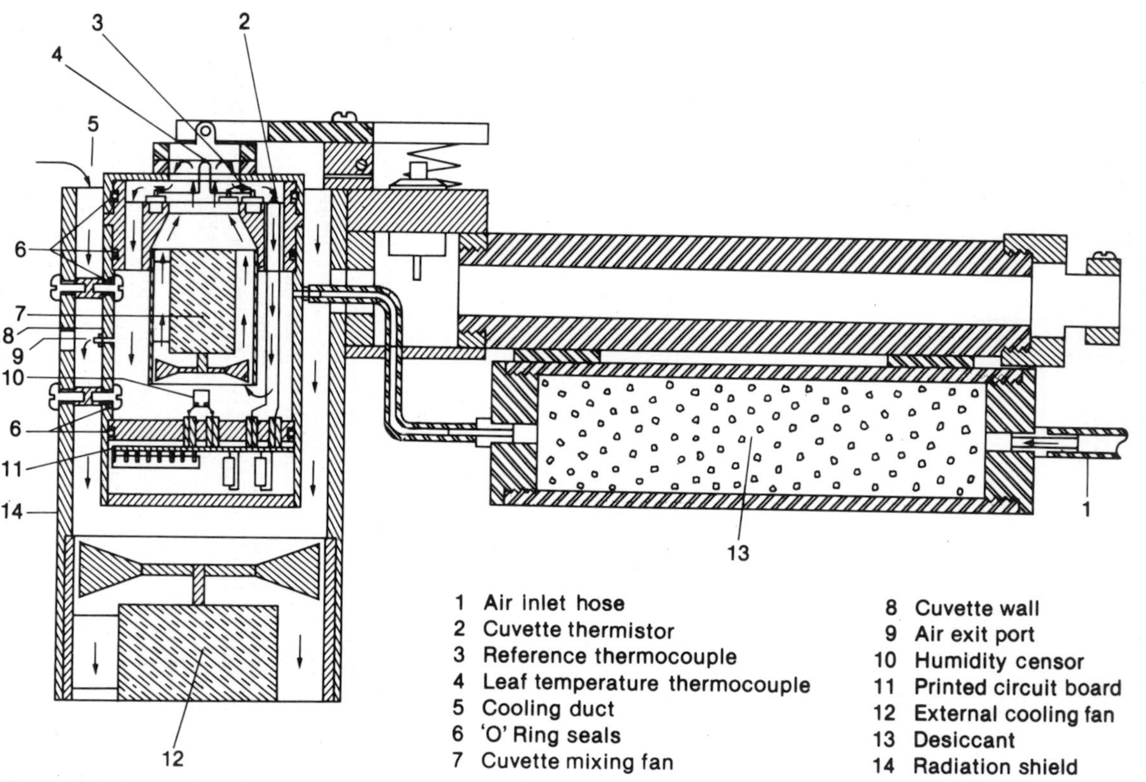

Figure 8.2 Sensor head of the LI-1600 steady-state diffusion porometer (courtesy of LI-COR inc., Lincoln, Nebraska, USA).

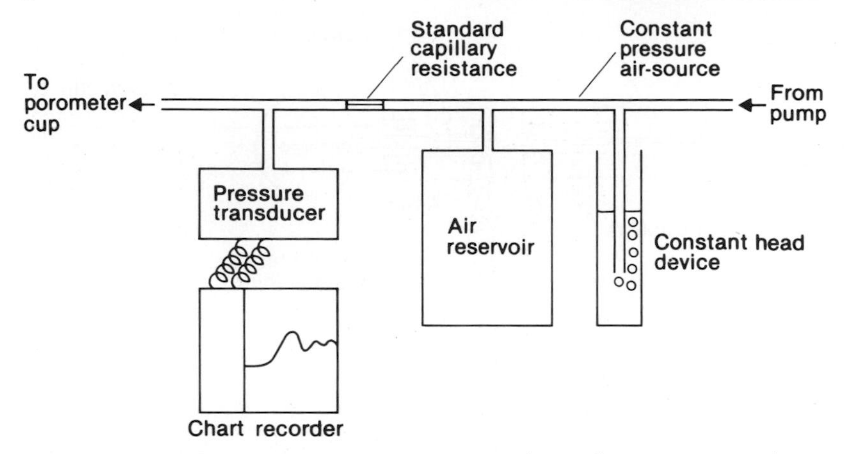

Figure 8.3 An automated Gregory and Pearse viscous flow porometer. A pressure transducer and chart recorder have replaced the manometers in the earlier version.

appears reliable and popular and is suitable for work in the laboratory and field.

Diffusion porometers of this type are based upon a sensor element that changes its electrical resistance with the moisture content surrounding the element. The element can be positioned at a fixed distance from the leaf and water vapour diffuses out of the leaf onto an initially dry sensor at rates depending chiefly on the degree of stomatal opening. For increased accuracy temperature corrections are made and a temperature sensor is usually built into the leaf clamp/cup arrangement to facilitate the correction.

Recently, diffusion porometers of steady-state design have come onto the market and give direct read-out of leaf resistance, transpiration rates, leaf and chamber temperatures, chamber humidity, and photon flux density (Figs 8.1, 8.2).

Of the viscous flow type of porometers the Alvim porometer (Alvim 1965) is robust, compact and popular for field work.

For laboratory work a modified version of the Gregory and Pearse resistance porometer (Gregory and Pearse 1934) is ideal (Fig. 8.3). Essentially a constant pressure air-source is connected in series to a calibrated (standard) capillary resistance and a porometer cup. A pressure transducer connected to a chart recorder measures the pressure between the standard resistance and the porometer cup. Thus, the wider open the stomata, the lower the pressure measured. A constant pressure inflow of air is obtained with the use of a constant head device. Bubbles formed in the constant head device cause oscillations of pressure and these are smoothed out by a large reservoir connected to the system. Although it is advisable to

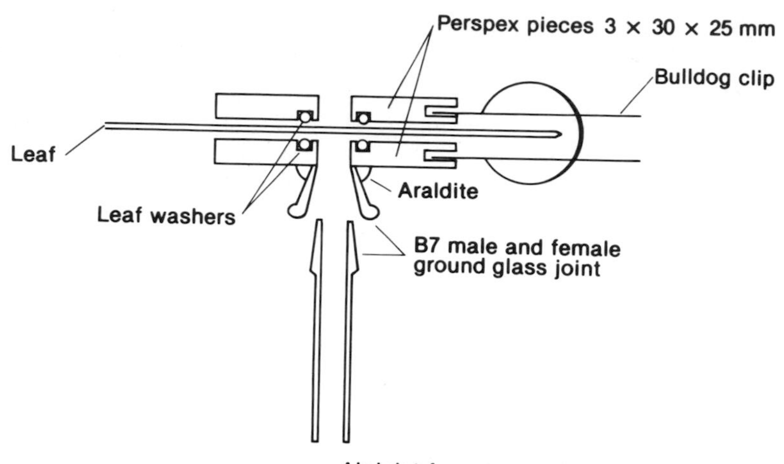

Figure 8.4 A simple detachable porometer clip (from Meidner 1965).

calibrate the capillary resistance so that correct units of leaf resistance (to mass flow) can be calculated stomatal movements can be observed on the chart recorder on a relative basis.

Numerous types of porometer cups have been designed. A simple but effective design is illustrated in Fig. 8.4. Although an air-tight seal between the cup and leaf are essential, damage to the leaf must not occur. To help to achieve this porometer cups with adjustable compression springs can be made or purchased. Porometer recordings can be automated and if lengthy experiments are being carried out with viscous flow porometers then a ventilated porometer cup permanently attached to the leaf, of the type developed by Spanner and Heath (1951), is desirable (Fig. 8.5).

One of the simplest ways of following stomatal movements is to measure transpiration rates (which, under most circumstances, are a reflection of stomatal aperture). Leaves, excised under water, are inserted into vials containing water or other test solution and the leaf plus vial periodically weighed. To prevent evaporative losses from the solution surface, paraffin oil may be layered on top. However, on occasions paraffin oil has been observed to penetrate the leaf tissue and, therefore, one alternative is to insert the leaf through a slit in parafilm which seals the top of the vial. At the end of the experiment the surface area of the leaf is measured and transpiration rates can be recorded in units of weight of water transpired/dm^2 leaf area/h.

It is also recommended that the experiment can be carried out in a growth room where temperature and light intensities are controlled.

If leafy twigs are used then a potometer-type of apparatus (Fig. 8.6) can be used to measure rates of water uptake and transpiration rates. The water uptake rate and transpiration rate may

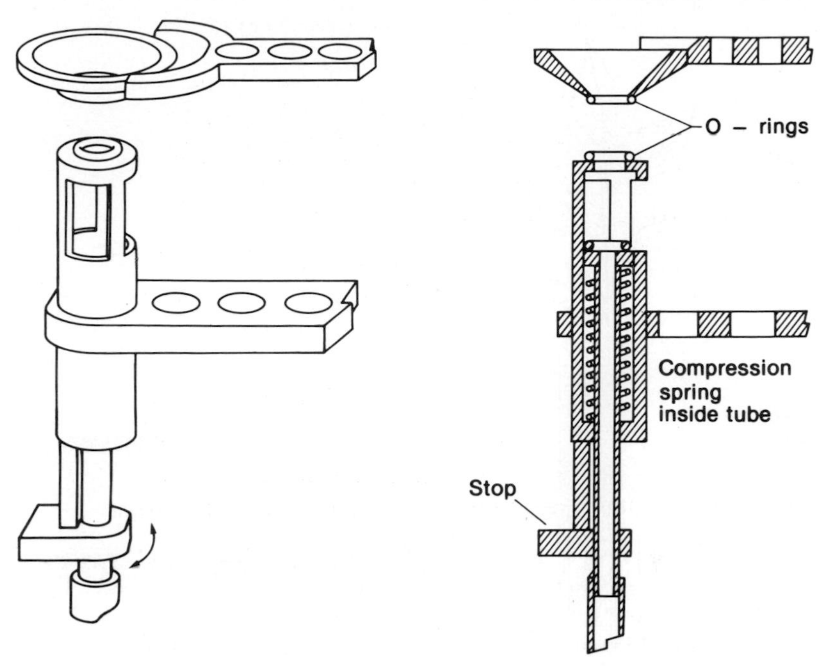

Figure 8.5 A porometer cup modified to permit prolonged attachment to a leaf for repeated measurements. Cut surfaces of the perspex upper area are sanded and either fire or solvent polished to maximise light transmission. The upper part of the cup is sloped to minimise shading and interference with air movement around the test area of the leaf. The lower part of the cup is shown in the ventilating position between measurements. When taking a measurement the inner tube is rotated in the lower arm to clear the stop, and by compression through the spring form and air-tight connection between the leaf in the cup and the pressurised air. The ventilating feature is adapted from Spanner and Heath, 1951 (from Bull. 809, Coll. Agric. Res. Centre, Washington State University).

be similar but not identical owing to absorption lags (themselves caused by capacitances and resistances in the tissues). Periodic weighing of the whole apparatus enables water vapour loss due to transpiration to be measured while the rate of movement of a meniscus along a graduated capillary tube is a measure of the water uptake rate by the shoot.

Thus, there are many alternative ways of monitoring stomatal behaviour and, depending on choice of material and measuring device, a host of experiments can be carried out. Using excised leaves or twigs the effects of metabolites and metabolic inhibitors on stomatal behaviour when applied via the transpiration stream can be

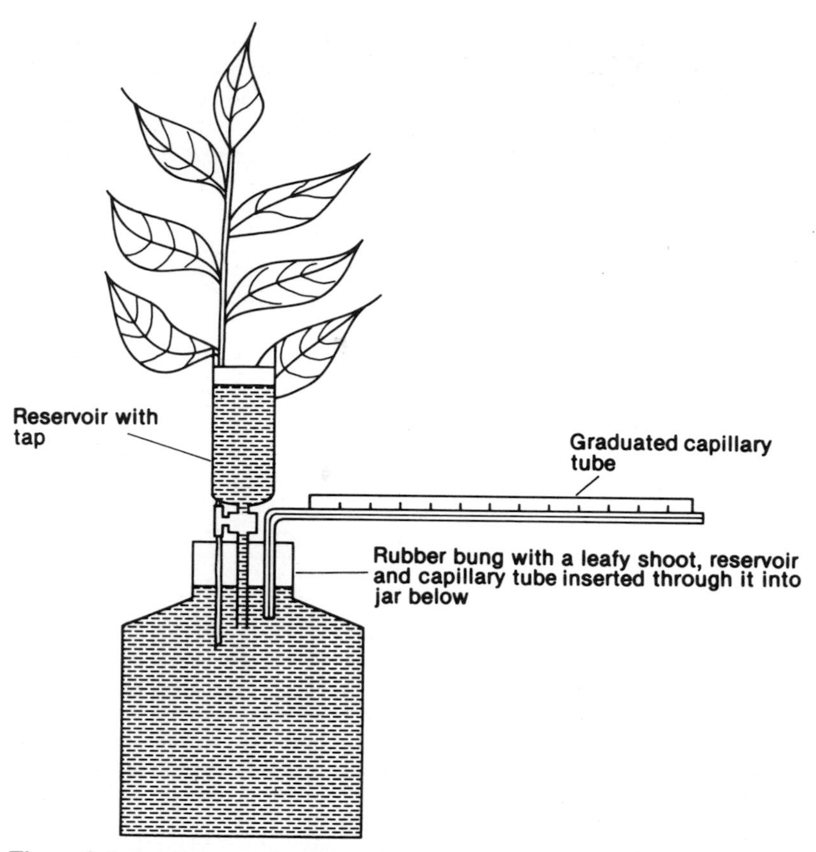

Figure 8.6 A potometer for measuring transpiration rates and water uptake rates.

determined. If a viscous flow porometer is used to measure the stomatal behaviour, leaves can be flushed with CO_2-free air or air of constant CO_2 concentration to determine if the metabolite is affecting stomata directly or indirectly by raising or lowering intracellular CO_2 levels. Also, viscous flow porometers allow the effects of various gases (e.g. N_2, CO_2-free air, ethylene) on stomata to be studied since the leaf air spaces can be flushed with these gases.

8.2 The use of epidermal strips for studying stomatal behaviour and operation

The direct effects of various environmental factors, metabolites and inhibitors on stomatal behaviour can be studied using epidermal strips.

The lower epidermis from *Vicia faba* or *Commelina communis* is recommended since more is known about stomatal responses in epidermal strips from these species than in others, the stomata are relatively large, and the epidermis readily peels from the leaves.

Growth of the plants is important if uniform and consistent results are to be obtained. The plants should always be well watered and grown under ideal conditions. The plants should also be used before flowering occurs.

If the experiment is to study an effect on the opening of stomata then plants should be placed in a darkened cupboard for 2–3 hours before the experiment to close the stomata.

Stomata of *Vicia faba* will not all close completely (even after many hours in the dark) while those of *Commelina communis* should. Results also tend to be more consistent with *Commelina communis*.

If the experiment is to study the effects of a particular factor on stomatal closing then plants are needed with wide open stomata. To achieve this plants can be enclosed in a plastic bag which is illuminated and CO_2-free air can also be pumped into the bag to stimulate wider and more rapid opening. However, one of the best ways to open stomata uniformly, rapidly and very wide is to float leaves on water with their undersurfaces down. The leaves should be illuminated (low intensity light is sufficient) and within 3 hours maximum opening occurs. Leaves can then be taken as required and the lower epidermis peeled from the leaves.

Fully expanded leaves three to four nodes from the apex of main shoots should be used. Some practice may be required to peel the epidermis from the leaves. Epidermal cells will be damaged in the peeling procedure but with *C. communis* this is less critical because subsidiary cells which surround the stomata remain intact and pressure effects of epidermal cells on the guard cells and stomatal movements are less evident. The epidermis should be floated on a 'holding solution' (0.1 mM $CaCl_2$), cuticle uppermost, until enough epidermis has been obtained for the experiment. The strips should then be cut into squares about 1 cm^2 with very sharp, fine-pointed scissors.

Incubating media: If *Commelina communis* is used, 75-100 mM KCl in 20 mM MES (2-[N-Morpholino] ethanesulphonic acid) buffer, pH 6.0, is recommended. If *Vicia faba* is used 1–10mM KCl in 10 mM MES buffer, pH 6.0, is recommended. Note that some extra K^+ will be added to the incubation medium when the MES buffer is adjusted to pH 6.0 with KOH.

The epidermal strips are transferred from the holding medium using a dissecting needle. The needle is lifted up beneath a piece of epidermis. The epidermis folds over the needle and floats off when the needle is dipped into the next solution. The epidermis is ideally incubated in the apparatus shown in Fig. 8.7. Temperatures are

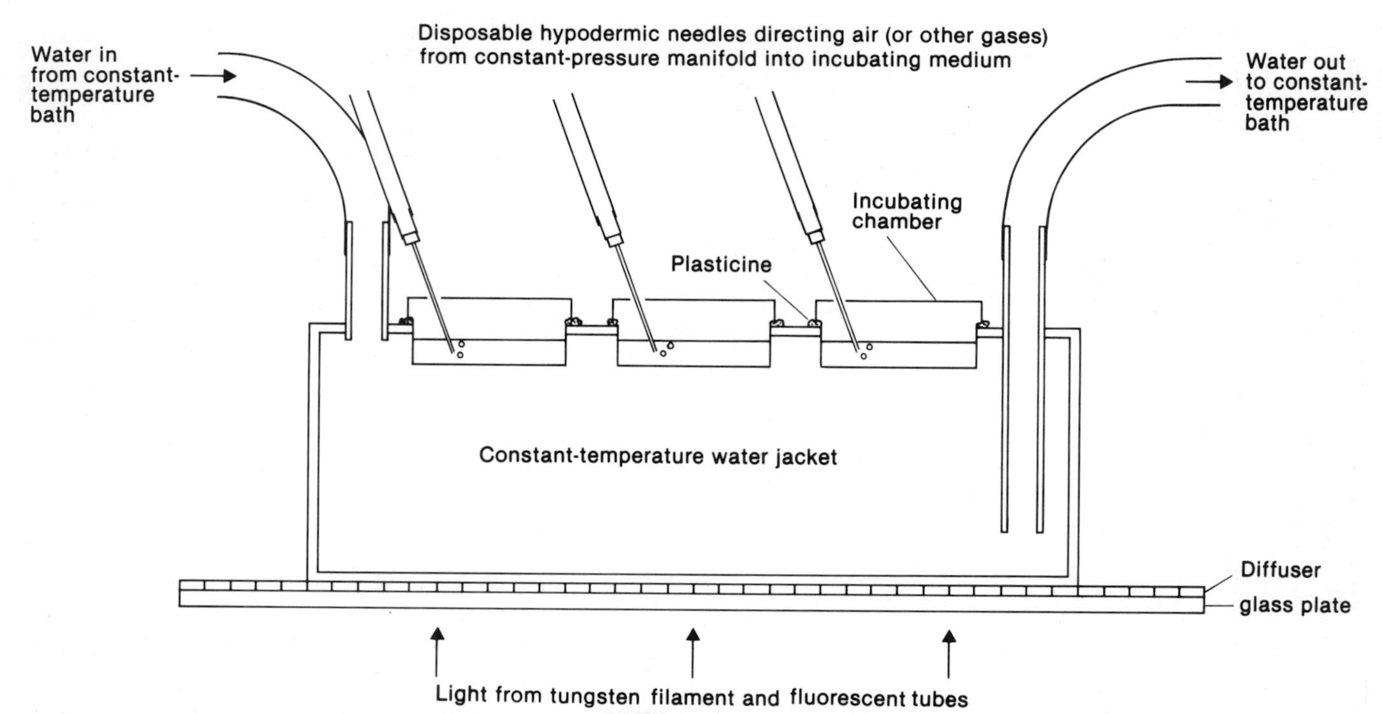

Figure 8.7 Apparatus used for incubating epidermal strips (Diagram courtesy of A. Plumbe.)

maintained constant by the perspex water jacket which is supplied with water from a constant temperature bath. Air, CO_2-free air or other gas can be bubbled through the medium via disposable syringe needles at a constant rate (about 50 cm^3/min). To obtain constant flow rate of gases simultaneously through a number of incubating chambers a large reservoir is needed which acts as a constant pressure device. The gas is pumped into the reservoir and flexible exit tubes are connected to the syringe needles. Lighting is preferably tungsten filament and fluorescent tubing positioned below the water jacket.

After the required incubation time (it takes about 1.5 hours for stomata to open fully in *C. communis* epidermis at 30°C) the epidermal strips are mounted in the incubation medium on a microscope slide with a coverslip and the aperture widths measured under a calibrated microscope with an eyepiece graticule (about × 400 is adequate). It is suggested that ten stomata are chosen at random (keeping away from the damaged edges of the epidermal strip) from each strip and their apertures measured. With experience, three pieces of epidermis can be examined in this way within 5 minutes. If a number of replicate experiments and/or different treatments are being conducted then the beginning of each incubation period must be staggered by 5 minutes to give sufficient time to measure the stomatal apertures at the end of the experiment.

Using this technique the effects of metabolites, metabolic inhibitors, light quality and quantity, temperature, different gases, etc. on stomatal behaviour can be monitored.

8.3 Histochemical tests on leaf epidermis with stomata open or closed

8.3a Potassium

The location of K^+ in epidermis of leaves with open or closed stomata can be determined histochemically using an adaptation of a method developed by Macallum (1905). In the test sodium cobaltinitrite forms the triple salt, sodim potassium cobaltinitrite, in the presence of K^+ which crystalise out at 'ice-cold' temperatures. For this reason all staining procedures which require 'ice-cold' temperatures should be carried out in a tray of crushed ice.

$$\underset{\text{sodium cobaltinitrite}}{Na_3Co(NO_2)_6 + 2K^+} \longrightarrow \underset{\substack{\text{sodium potassium cobaltinitrite} \\ \text{(yellow ppt.)}}}{NaK_2Co(NO_2)_6 + 2Na^+}$$

The yellow crystals which form are not easy to observe and so, after a suitable washing period, the triple salt is reacted with ammonium sulphide to form a black precipitate of cobalt sulphide.

Procedure: 1. Epidermis with open or closed stomata (see §8.2 to obtain leaves with open or closed stomata) is rinsed for 1–2 minutes in 'ice-cold' distilled water to rid the tissue of K^+ from broken cells (either epidermal or adhering mesophyll cells). Twenty mM calcium sulphate or calcium chloride has also been used as a washing medium with good results. The Ca^{2+} exchanges for K^+ situated in the cell walls and cell debris. The calcium salt is then rinsed out of the tissue by immersing the tissue in 'ice-cold' distilled water for 2 minutes. The tissue should be agitated at all stages of the procedure so that air bubbles do not insulate the cells from contact with the incubation medium.

2. The tissue is transferred to 'ice-cold', freshly-made sodium cobaltinitrite solution (the solution lasts for at least 2 weeks when kept in the refrigerator). Sodium cobaltinitrite solution made from commercially available solid does not always give good results and sodium cobaltinitrite solution prepared as follows is recommended: add 75 ml 13 per cent acetic acid to 20 g cobalt nitrate and 35 g sodium nitrite. Poisonous fumes of nitrogen dixoxide are evolved and so the manipulation should be carried out in a fume cupboard, preferably with a stream of air bubbling through the solution to speed up evolution of the fumes. For most types of epidermal tissue a 10-minute exposure to the sodium cobaltinitrite is sufficient to give good results. However, with some tissues, particularly heavily cuticularised epidermis, longer periods of incubation up to 20 minutes may be necessary.

3. The tissue is now washed in 'ice-cold' distilled water until no more yellow stain flows from the tissue. This usually entails washing the tissue for about 2 minutes in three changes of ice-cold distilled water.

4. The tissue is next immersed in freshly-made 5 per cent yellow ammonium sulphide solution at room temperature for about 2 minutes. This step must be carried out in a fume cupboard.

5. Finally, the tissue is briefly washed in water to get rid of surplus ammonium sulphide and mounted on a slide in a drop of water. Permanent mounts can be made but there does appear to be some slight loss of intensity of staining with time.

The procedures employed will close open stomata. The pattern of K^+ localisation obtained with the histochemical technique compares favourably with that obtained with the electron microprobe and it is considered that the histochemical test gives a reasonably accurate picture of the true K^+ location. Freeze-dried tissue does not give good results, unlike that apparently observed in animal tissue. It is also considered that the test is specific for K^+.

Some results obtained using the procedures described above are shown in Figs. 7.1–7.6.

8.3b Starch

Leaf material with open or closed stomata is obtained in the manner described in §8.2. Epidermal strips are peeled from the leaves (or paradermal sections made if the epidermis does not readily detach from the leaf) and the starch test is carried out by placing the tissue in a dish of iodine/potassium iodide solution (0.3 g iodine plus 1.5 g potassium iodide in 100 cm^3 distilled water). The tisue is left in this solution for about 5 minutes, and the excess stain then rinsed out with distilled water. The tissue is mounted in water on a slide with a cover-slip and the extent of blackening (a blue-black colouration is a positive test for starch) of the chloroplasts (the location of the starch) within the guard cells is assessed by a scoring technique. The intensity of blackening is gauged from 0 (no blackening) to 5 (most intense blackening). A set of previously prepared photographic standards will assist in the scoring.

Using this technique it is possible to correlate stomatal aperture with guard cell starch levels and to establish that starch is present in guard cells of almost all species, including *Paphiopedilum* (lady slipper orchid) species which lack *chloroplasts*, but is absent in the *Allium* genus.

8.4 Determination of the osmotic potential of guard cells when stomata are open or closed

The incipient or limiting plasmolysis technique can be used to measure osmotic potentials of guard cells. However, the reader must be aware that there are faults inherent in the technique (*see* §4.1) which will tend to cause underestimates of the true values. The concentration of the plasmolyticum in which 50 per cent of the cells show plasmolysis is determined. When 50 per cent of the guard cells are plasmolysed the osmotic potential of the external plasmolyticum is equal to the mean value for the vacuolar sap of the cells.

The technique is as follows: A range of sucrose concentrations from 0.1 to 1.0 M is prepared. (Mannitol is often used as a plasmolyticum because it is considered less likely to be metabolised than some other plasmolytica such as sucrose. However, mannitol is less soluble than sucrose and the higher concentrations needed in this experiment are unattainable. Sucrose has the advantage of being a larger molecule than mannitol and will therefore penetrate the cells more slowly, i.e., the reflection coefficient of sucrose is greater than mannitol.) To aid in observing the plasmolysis the sucrose solution can be prepared in 0.01 per cent neutral red which has a negligible osmotic effect.

Leaves with open or closed stomata are obtained in the manner described in §8.2. The epidermis is peeled from the leaves and the apertures of a representative sample of stomata measured.

A piece of the epidermis is floated, cuticle uppermost, on one concentration of the plasmolyticum. The epidermis is agitated and any bubbles attached to the tissue are dislodged. After 10 minutes the epidermal strip is mounted in the plasmolyticum on a slide with a cover slip and the number of plasmolysed and non-plasmolysed guard cells counted. This observation is made as quickly as possible (within 5 minutes) and then the procedure is repeated for each of the sucrose concentrations and for open and closed stomata. It is important to keep the plasmolysing time as constant as possible for each of the sucrose concentrations (see Willmer and Beattie 1978).

The percentage plasmolysis for each sucrose concentration is plotted against sucrose concentration (M) and the point of 50 per cent plasmolysis is equivalent to the guard cell osmotic potential. The molar concentration of sucrose can be converted to osmotic potential from Table 8.1.

Table 8.1 The relationship between molar sucrose concentration and osmotic potential at 20° C

Sucrose concentration (M)	Osmotic potential (bar)
0.1	−2.6
0.2	−5.2
0.3	−8.0
0.4	−11.0
0.5	−14.1
0.6	−17.6
0.7	−21.2
0.8	−25.2
0.9	−29.3
1.0	−34.1

Guard cells of all species are not ideal material for observing plasmolysis. Guard cells of *C. communis*, however, behave reasonably well although typical concave or convex plasmolysis with rounding off of the protoplast does not always occur.

Because of the problems encountered in making measurements of guard cell dimensions when stomata are open and closed the errors in the osmotic potential values due to volume changes of the cells are generally ignored.

Further reading and references

Alvim, P. de T. (1965) A new type of porometer for measuring stomatal opening and its use in irrigation studies, *UNESCO Arid Zone Res.*, **25**, 325–9.

Beardsell, M. F., Jarvis, P. G. and Davidson, B. (1972) A null-balance diffusion porometer suitable for use with leaves of many shapes, *J. Appl. Ecol.*, **9**, 677–90.

Gregory, F. G. and Pearse, H. L. (1934) The resistance porometer and its application to the study of stomatal movement, *Proc. Roy. Soc.*, *B.*, **114**, 477–93.

Kanemasu, E. T., Thurtell, G. W. and Tanner, C. B. (1969) Design, calibration and field use of a stomatal diffusion porometer, *Plant Physiol.*, **44**, 881–5.

Lloyd, F. E. (1908) The physiology of stomata, *Publ. Carnegie Inst.*, *Wash.*, **82**, 1–142.

MaCallum, A. B. (1905) On the distribution of potassium in animal and vegetable cells, *J. Physiol.*, **32**, 95–118.

Meidner, H. (1965) A simple porometer for measuring the resistance to air flow offered by stomata, *School Sci. Rev.*, **47**, 149–51.

Meidner, H. and Mansfield, T. A. (1968) *Physiology of Stomata*. McGraw-Hill, London.

Slatyer, R. O. and Jarvis, P. G. (1966) Gaseous diffusion porometer for continuous measurement of diffusive resistance of leaves, *Science*, **151**, 574–6.

Spanner, D. C. and Heath, O. V. S. (1951) Experimental studies of the relation between carbon assimilation and stomatal movement, *Ann. Bot.*, *NS*, **XV**, 319–431.

Washington State University (1975) Measurement of stomatal aperture and diffusive resistance. Bull. 809, 1–40. *Pub., Coll. Agric. Res. Centre, Washington State University*, *USA*.

Willmer, C. M. and Beattie, L. N. (1978) Cellular osmotic phenomena during stomatal movements of *Commelina communis*. 1. Limitations of the incipient plasmolysis technique for determining osmotic pressures, *Protoplasma*, **95**, 321–332.

Index

84-4400 MCL

QK
873
.W54
1983

Willmer, Colin M.

Stomata

DATE DUE

FEB 25 '88			
FEB 22 '88			

84-4400 MCL

MILTON CAMPUS LRC

PENSACOLA JUNIOR COLLEGE
MILTON CAMPUS LIBRARY